IMPROVING WATER, SANITATION, AND HYGIENE IN SCHOOLS

A GUIDE FOR PRACTITIONERS AND POLICY MAKERS IN MONGOLIA

Kevin Tayler and Asako Maramuya

APRIL 2020

ASIAN DEVELOPMENT BANK

ADB

© 2020 Asian Development Bank
6 ADB Avenue, Mandaluyong City, 1550 Metro Manila, Philippines
Tel +63 2 8632 4444; Fax +63 2 8636 2444
www.adb.org

Some rights reserved. Published in 2020.

ISBN 978-92-9262-170-4 (print), 978-92-9262-171-1 (electronic), 978-92-9262-172-8 (ebook)
Publication Stock No. TIM200125-2
DOI: http://dx.doi.org/10.22617/TIM200125-2

The views expressed in this publication are those of the authors and do not necessarily reflect the views and policies of the Asian Development Bank (ADB) or its Board of Governors or the governments they represent.

ADB does not guarantee the accuracy of the data included in this publication and accepts no responsibility for any consequence of their use. The mention of specific companies or products of manufacturers does not imply that they are endorsed or recommended by ADB in preference to others of a similar nature that are not mentioned.

By making any designation of or reference to a particular territory or geographic area, or by using the term "country" in this document, ADB does not intend to make any judgments as to the legal or other status of any territory or area.

Please contact pubsmarketing@adb.org if you have questions or comments with respect to content, or if you wish to obtain copyright permission for your intended use that does not fall within these terms, or for permission to use the ADB logo.

Corrigenda to ADB publications may be found at http://www.adb.org/publications/corrigenda.

Note:
In this publication, "$" refers to United States dollars.

Cover design by Gato Borerro.

Contents

Tables, Figures, and Boxes

Boxes

Acknowledgments

This volume, *Improving Water, Sanitation, and Hygiene in Schools: A Guide for Practitioners and Policy Makers in Mongolia,* is part of collaborative efforts from the Government of Mongolia and the Asian Development Bank (ADB) to improve water, sanitation, and hygiene (WASH) in schools in Mongolia. It was first conceived while designing a grant assistance to the Government of Mongolia from the Japan Fund for Poverty Reduction in 2016–2019 to improve the school dormitory environment for primary students in one of the poorest regions in Mongolia.

This publication was written by Kevin Tayler, civil engineer, and consultant, and Asako Maruyama, senior education specialist, ADB under the overall guidance of James P. Lynch, director general; M. Teresa Kho, deputy director general; Ying Qian, advisor; and Sangay Penjor, director, East Asia Department. Antoine Morel, principal environment specialist, Southeast Asia Department, contributed to initial phases in the preparation of this publication. Ruth D. Benigno, associate project analyst, coordinated the publication, and Edith Creus, consultant, did the layout and typesetting.

The team benefited from valuable comments during the peer review process. The peer reviewers were Amy Leung, former director general, Sustainable Development and Climate Change Department and East Asia Department; Norio Saito, director, South Asia Department; Satoshi Ishii, principal urban development specialist, and Alan Baird, principal urban development specialist, Southeast Asia Department; Karina Veal, former senior education specialist, Sustainable Development and Climate Change Department; and Tuul Badarch, senior project officer (infrastructure), Mongolia Resident Mission.

Abbreviations

ADB	–	Asian Development Bank
CHAST	–	Children's Hygiene and Sanitation Training
IRC	–	International Water and Sanitation Centre
KAP	–	knowledge, attitudes, and practices
MBBR	–	moving bed biological reactor
MECSS	–	Ministry of Education, Culture, Science and Sports
O&M	–	operation and maintenance
PEX	–	cross-linked polyethylene
PHAST	–	Participatory Hygiene and Sanitation Transformation
SATO	–	safe toilet
TDS	–	total dissolved solids
UDDT	–	urine-diverting dry toilet
UNICEF	–	United Nations Children's Fund
uPVC	–	unplasticized polyvinyl chloride
USPW	–	uniform series present worth
VIP	–	ventilated improved pit
WASH	–	water, sanitation, and hygiene
WHO	–	World Health Organization

Weights and Measures

dGH	–	degrees of hardness
kg/m^3	–	kilogram per cubic meter
km	–	kilometer
kW	–	kilowatt
kWh	–	kilowatt-hour
$kWh/m^2/day$	–	kilowatt per meter squared per hour
LPCD	–	liter per capita per day
l/c.yr	–	liter per capita per year
l/sec	–	liter per second
m	–	meter
mm	–	millimeter
m/s^2	–	meter per second squared
m^3/h	–	cubic meter per hour
MW	–	megawatt
mg/l	–	milligram per liter
ml	–	milliliter
mg-eq/liter	–	milligram-equivalents per liter
m^3	–	cubic meter
μg/L	–	microgram per liter

Executive Summary

Good water, sanitation, and hygiene (WASH) services protect health and environment, and provide users with convenience and privacy. In schools, WASH is particularly important because children who are taught about better hygiene and have access to the requisite facilities to practice what they are taught develop hygiene habits, which they will follow throughout their lives. Adequate water and sanitation facilities also have secondary benefits—children who learn in a clean and hygienic environment perform better because they are healthier and less likely to be absent from school; and girls are more likely to attend school and remain in education if they have access to appropriate sanitation facilities that are separated from those used by boys to provide privacy.

In Mongolia, especially in rural areas, WASH in schools has been given low priority for a long time. Beginning around 2010, poor conditions of WASH facilities and services in schools started to draw the attention of policy makers, which eventually led to the Ministry of Education, Culture, Science and Sports; the Ministry of Health; and the Ministry of Finance jointly issuing the minimum requirements for WASH in schools in 2015.

Meeting the minimum requirements for WASH presents a considerable challenge to many rural schools, which account for more than half of all schools in Mongolia. One of the major constraints on school WASH improvements is the absence of an easily accessible water source or water availability, which makes it difficult to devise affordable WASH schemes. Even if a viable water source exists near schools, providing sustainable WASH improvements at affordable costs remains a challenge in rural areas. This is because Mongolia's harsh winter, with temperatures falling below freezing point for long periods, requires WASH facilities to be heated to prevent freezing. Water supply and sewer pipes laid close to the ground surface must be heated. In the absence of centralized heating systems, which is mostly the case in many rural schools, heating costs are high.

Rural schools in Mongolia often rely on stand-alone, basic WASH schemes because settlement-wide water supply, sewer, and heating systems are not available. Constant rural to urban migration reduces population growth in rural areas, leading to small population sizes, with many families continuing to live seminomadic lifestyles. The small populations and the large distances between settlements in rural areas increase the cost of WASH services, and cause significant logistical difficulties in sustaining those services.

Although there are existing reference materials that provide useful and generally applicable guidance on the improvement of school WASH facilities, they hardly address the specific physical and institutional challenges presented by the harsh climate and the small size and isolation of rural settlements in Mongolia. As a result, technologies selected for WASH facilities in schools are often incompatible with existing infrastructure and services, climate conditions, and specific needs of schools. This volume, *Improving Water, Sanitation, and Hygiene in Schools: A Guide for Practitioners and Policy Makers in Mongolia* (henceforth, the Guide), seeks to fill the gap by providing guidance on the processes to be followed and the decisions required to improve WASH in schools in Mongolia.

This Guide is intended for use in schools, particularly by school management, teachers, social workers, and school doctors; national and local policy makers and administrators; and engineers, architects, and cost estimators involved in designing and constructing school WASH facilities. The main sections of the Guide provide essential information for planning, implementing, and managing improved WASH in schools, covering legislation, standards, and norms on school WASH in Mongolia (section II); steps in planning improved WASH in schools (section III); roles and responsibilities, and tasks involved in implementing and managing improved WASH in schools (section IV); and health impacts of WASH and hygiene education (section V). The appendixes are designed to serve as technical and practical reference materials. They include more information on climate conditions, water availability, and quality in Mongolia (Appendix 1); process and tools for assessing WASH needs in schools (Appendix 2); design and technological options for water supply in cold climates (Appendix 3); design and technological options for sanitation in cold climates (Appendix 4); hygiene education approaches (Appendix 5); and methodology for estimating and comparing costs of improved WASH scheme options (Appendix 6).

The lack of physical, institutional, and financial resources makes it difficult and costly for many rural schools to improve WASH services, even to meet the minimum requirements set by the Government of Mongolia. This Guide suggests options for incremental improvements, designed to bring about immediate benefits, while paving the way for eventual full compliance with the minimum requirements and higher standards.

The minimum requirements for WASH in schools specify that each school should have indoor water-flushed toilets, and where necessary, outdoor latrines with adequate insulation, ventilation, lighting, and handwashing facilities located at least 20 meters away from school and dormitory buildings. However, many rural schools continue to build outdoor toilet blocks without any enclosed and lighted access way and handwashing facilities as main sanitation facilities in schools. This is largely because these schools have no easy access to viable water sources, public sewerage and wastewater treatment facilities, and settlement-wide heating systems. These basic outdoor toilet blocks, therefore, need to be retained in the short term, while efforts to replace them with hygienic toilets grouped in sanitation blocks that meet the minimum requirements should be made in the longer terms. Some aspects of the minimum requirements could be revised, if improved forms of dry direct-drop toilets become available, to allow outdoor toilet blocks to be located closer to school buildings.

At the *aimag* level, there is a need to develop capacity to gather and organize information, identify priorities, and support school WASH improvement exercises in coordination with departments and organizations involved in planning and management of WASH services. For instance, systems should be developed at the *aimag* level to provide services such as periodic desludging of septic tanks, and support schools in procuring repair services and spare parts of WASH-related tools, equipment, and facilities, which are beyond the capacity of individual schools.

At the school level, systems need to be in place to ensure continued operation, maintenance, and minor repair of WASH facilities. This would require rewriting of job descriptions, employment of additional staff to undertake operation and maintenance tasks, and contracting private sector or community organizations to provide operation, maintenance, and repair services.

Good sanitation and water treatment facilities alone cannot prevent transmission of pathogens. Proper hygiene practices, especially, handwashing with soap are essential to block transmission. Therefore, hygiene education, aiming to deliver a small number of key messages, should be part of WASH planning and improvement exercises. Support systems should be established at the *aimag* level to promote hygiene education activities at schools.

More adequate funding is necessary for ongoing WASH-related activities, including operation, maintenance, and repair of WASH facilities, and hygiene education activities. School budgets for WASH-related activities should be increased. It is recommended that the government issue guidelines requiring *aimag* and Ulaanbaatar education departments to allocate more funds for operation, maintenance, and repair of school WASH facilities and hygiene education activities.

I. Introduction

Water, sanitation, and hygiene (WASH) is one among many issues that vie for the attention of decision makers. Why should WASH be given priority? At a general level, the United Nations Sustainable Development Goal 6 is to "ensure availability and sustainable management of water and sanitation for all." Targets associated with Goal 6 refer to the need to achieve access to safe and affordable drinking water and adequate and equitable sanitation and hygiene for all by 2030.[1]

Good WASH services are important because of their role in protecting health and environment, and providing users with greater convenience and privacy. According to the United Nations Children's Fund (UNICEF), children who are denied access to safe water and convenient sanitation facilities are exposed to diarrheal diseases and respiratory infections, which deprive them of nutrients, leading to stunted growth, illness, and frequent absence from school classes (UNICEF 2013). In developing countries, fecal–oral diseases, including various forms of dysentery, cholera, and typhoid, are a major cause of mortality among children under 5 years of age. They also contribute to morbidity among older children and adults, reducing their ability to study and work, and leading to poor education outcomes and work performance. Good water supply and sanitation facilities and services help to prevent transmission of fecal–oral diseases—diseases that pass from one person to another when a second person ingests a pathogen (disease-causing organism) excreted by an infected person.

In schools, WASH is particularly important because children who are taught about better hygiene and have access to the requisite facilities to practice what they are taught are likely to develop hygienic habits, which they will follow throughout their lives. They can also become change agents, communicating messages about good hygiene to siblings, parents, and friends, and promoting demand for improved water and sanitation facilities.

Adequate water and sanitation facilities have further benefits. Evidence shows that children who learn in a clean and hygienic environment perform better because they are healthier and are less likely to be absent from school. From a gender perspective, girls are more likely to attend school and remain in education if they have access to good sanitation facilities that are separated from those used by boys to provide privacy.

A. Water, Sanitation, and Hygiene Conditions in Schools in Mongolia

Almost half of Mongolian students enroll at schools in Ulaanbaatar, which account for one-third of Mongolian schools. The majority of the schools in Ulaanbaatar are connected to water supply and sewered sanitation. The

---

[1]   See UN Water website at http://www.unwater.org/sdgs/a-dedicated-water-goal/en/.

other half of students attend schools either in *aimag* centers, which are relatively well served by urban WASH services; or *soum* centers, and *baghs*, which have no access to piped water supply and sewerage services.

Uneven population distribution, small size of rural settlements, and large distances between those settlements often result in poor WASH provision, particularly in isolated *soum* centers and *baghs*, which have limited access to back-up advice and guidance on technical and operational matters. Problems during WASH planning and implementation such as deviations from engineering drawings, and the use of poor-quality equipment and materials, are prone to occur especially in isolated *soum* centers and *baghs* without adequate supervision.

Rural schools face particular challenges in accessing water. A study on school dormitory conditions conducted by the Ministry of Education, Culture, Science and Sports (MECSS) in 2007 revealed that

(i)   water for 74% of 502 dormitories then in use was carried from outside water kiosks and wells;

(ii)   46% of water supplied to schools did not meet hygienic standards; and

(iii) no less than 78% of schools only had outdoor latrines, most of which were unsafe and unhygienic.

The then Australian Agency for International Development (AusAID) found that three of the six schools they visited had recently constructed buildings with internal toilets, plumbing, and handwashing stations, but had no water supply. Consequently, the indoor facilities were used as storage space, while students continued to use inadequate outdoor toilets (AusAID 2011).

Bathing and water-flushed sanitation facilities in many rural schools are nonoperational because there is no water to supply them; and heating systems for shallow water pipes have proved to be inadequate, leading to freezing of the pipe contents and pipe fractures.

In some *soum* centers, water supply systems that rely on district heating systems installed before Mongolia started to transition from a centrally planned to a market economy have failed, because the district heating systems are no longer operational.

B. Challenges in Improving Water, Sanitation, and Hygiene in Schools in Mongolia

Over half the country's population resides in Ulaanbaatar, with the remainder thinly spread across Mongolia's large land area. Constant rural to urban migration reduces population in rural areas, leading to small populations in rural *soum* centers, with many of their inhabitants continuing to follow seminomadic lifestyles. Small populations of rural *soum* centers and large distances between rural settlements increase the cost of WASH services and present significant logistical difficulties in sustaining those services.

Mongolia experiences harsh winters, with temperatures falling to around –50°C in winter. This has profound implications for the design of WASH facilities since water in pipes and tanks freezes, causing system failure, unless the pipes and tanks are adequately protected. In cities and towns with district heating systems, pipes can be heated by laying them along heating ducts. Where there is no district heating, pipes can be heated with electrically powered heating coils. However, this option is expensive and is dependent on the existence of a reliable 24-7 power supply. The other option, which is also expensive, is to bury pipes below the lowest depth to which frost penetrates. These examples highlight the importance of ensuring that technologies adopted are suitable for use in extreme cold climates.

Water availability may be an issue for rural schools, which makes it difficult to devise affordable WASH schemes. Identification of affordable technologies that are compatible with local physical realities, including harsh winter conditions and lack of good water sources, presents a particular challenge. In some areas, problems are compounded by the poor quality of available water sources (for more information on the physical context of Mongolia that poses a challenge in improving water and sanitation facilities, see Appendix 1). Taken together with the small size of most rural settlements, these factors help to explain few rural schools have access to good WASH services.

School WASH facilities provide the interface with the intended users—students, teachers, and others who use the school premises. However, they are only one part of overall water supply and sanitation systems. If the other components of the systems are either absent or inadequate, school-level facilities will remain unused, which is not uncommon in rural schools in Mongolia.

Limited availability of engineers with knowledge and experience of technologies suitable for use in small rural settlements is another challenge. Certified engineers are usually trained in conventional engineering practices, focusing on solutions that are appropriate for large urban centers. Those solutions, however, are unlikely to be feasible in many rural areas, where water resources are poor and capacity to manage and maintain WASH systems is limited.

Schools also face difficulties in ensuring the availability of resources, including funding, which are required for successful operation and maintenance (O&M) of WASH facilities.

The lack of physical, institutional, and financial resources in many rural Mongolian schools means that improving WASH services, even to meet the minimum requirements, specified by the government, is both difficult and costly. In view of this, this volume, *Improving Water, Sanitation, and Hygiene in Schools: A Guide for Practitioners and Policy Makers in Mongolia* (henceforth, the Guide) recognizes a need for incremental improvements designed to bring about immediate benefits, while paving the way for eventual full compliance with the minimum requirements and higher standards.

C. Need for Guide on Improving Water, Sanitation, and Hygiene in Schools in Mongolia

Although existing reference materials (Adams et al. 2009, UNICEF and IRC 1998, UNICEF and GIZ 2013) provide useful and generally applicable guidance on school WASH facilities, they hardly address the specific physical and institutional challenges presented by Mongolia's harsh climate and the small size and isolation of its rural settlements. As a result, technologies selected for WASH facilities in schools are often incompatible with existing infrastructure and services, climate conditions, and specific needs of schools. This Guide responds to this situation by providing guidance that covers both generic aspects of school WASH planning, and specific technical challenges posed by Mongolia's unusual physical and demographic conditions. In particular, it describes a step-wise procedure for assessing needs; identifying options for addressing those needs; developing, costing, implementing, and operating WASH schemes; and designing hygiene education activities.

The Guide focuses on technologies that meet Mongolia's regulatory environments, that are appropriate for use in small isolated settlements, and that will continue to function during Mongolia's harsh winters. It emphasizes the importance of ensuring that operational and management needs of selected technologies can be met in remote rural areas where more than half of Mongolian schools are located.

The focus throughout the Guide is on effective service provision through sustainable WASH facilities. This requires effective management systems, including appropriate arrangements for continued operation, maintenance, and repair of WASH facilities. Guidance is also provided on hygiene education. Because sustainable WASH service provision depends on proper use of WASH facilities, the guidance on hygiene education includes a strong focus on educating users in the correct use of facilities.

D. Target Users

The target users of this Guide include

(i)   schools, particularly school management, which initiates school WASH improvements, and operates and manages school WASH facilities;

(ii)   government agencies at the national, *aimag, soum,* and *bagh* levels, which are involved in planning, improving, and managing WASH in schools;

(iii)   engineers, architects, and cost estimators responsible for designing and constructing school buildings and WASH facilities;

(iv)   teachers, social workers, and school doctors who are responsible for various aspects of WASH service delivery and hygiene education in schools; and

(v)   policy makers and administrators at the national level with responsibility for setting and enforcing standards for WASH in schools.

II. Legislation, Standards, and Norms on Water, Sanitation, and Hygiene in Schools

The water, sanitation, and hygiene (WASH) facilities and services in schools form part of the country's water supply and sanitation systems, which are subject to binding and guiding rules.

A. National Laws and Standards

The central piece of legislation governing the water sector is the Water Law of 2004, as amended in 2005 and 2009. Other laws with relevance to WASH in schools include the following:

(i)   Law on Urban Water Supply, Sanitation and Sewerage, 2002 (updated in 2011);

(ii) Law on Springs, 2003;

(iii) Law on Water Transportation, 2003;

(iv) Law on Sanitation, 1998 (updated in 2001); and

(v)   The Environmental Protection Law, 1995 (updated in 2012), designating the leading authority to coordinate various stakeholders involved in protecting the environment, developing and adopting standards, and ensuring their implementation.

The standards that have a bearing on WASH in schools are as follows:

(i)   MNS494300–1980: standards to be achieved relating to water quality and wastewater discharges;

(ii) MNS5924–2008: standards and technical requirements for toilets and sewage pits; and

(iii) MNS5924–2015: standards and technical requirements in relation to on-site sanitation (*Pit Latrine and Sewage Pit. Technical Requirements*).

B. Minimum Requirements for Water, Sanitation, and Hygiene in Schools

In 2015, the Ministry of Education, Culture, Science and Sports (MECSS), together with the Ministry of Health and Ministry of Finance, issued the minimum requirements for WASH in schools (MECSS, Ministry of Health, and Ministry of Finance 2015). These cover (i) location of facilities, particularly sanitation facilities; (ii) number of users per facility; (iii) drinking water quality; (iv) per capita water requirements; (v) provision for small children and the disabled; (vi) provision for adolescent girls; and (vii) hygiene education. Each requirement is discussed in detail in this section.

1. Location of Facilities

Different types of WASH facilities should be in the following locations:

(i)   **Reliable water points.** These should be available in or close to toilet blocks and kitchens and provided with soap. These water points should be accessible to staff and children with disabilities.

(ii) **Toilets for schools and kindergartens.** These must be indoors.

(iii) **Dry on-site sanitation facilities.** These should be located at least 20 meters (m) from a dormitory block or communal building (including classrooms), 25 m from water distribution kiosks and deep water wells, and at least 200 m from other water sources, including rivers and streams.

(iv) **Outside toilets.** These should be accessible in all weather conditions and children should not have to walk more than 30 m to an outside toilet. Outdoor latrines must have insulation, ventilation, and lighting; must be lockable from the inside; and be equipped with a waste bin with lid.

(v)   **Laundry facilities.** These should include soap or detergent and hot water, and should be available in all schools with dormitories.

(vi) **A functional drainage system or soak pit.** These should be provided at each cleaning and laundry point.

The implication of the requirement that toilets serving schools and kindergartens must be indoors and the prohibition on locating dry sanitation facilities within 20 m of dormitory blocks and communal buildings is that flush toilets are the only conventional option allowed for schools and kindergartens. This is extremely challenging if schools or kindergartens have limited access to water.

Where outdoor latrines serve dormitories, the minimum requirements for WASH in schools stipulate that washing facilities must be indoors. Where water supply to schools is constrained, it is theoretically possible to site waterless toilets in separate buildings at least 20 m from other buildings. One challenge then will be to design a toilet block that does not smell, and with access for periodic removal of accumulated excreta (section III and Appendix 4 provide further information).

2. Users per Facility

Each facility should be provided for the number of users specified as follows:

(i)   **One shower for 20 users in a school with dormitories.** The showers should be separate for girls and boys, students, and residential staff. At least one shower should be adapted for use by females with disabilities and one by males with disabilities.

(ii) **Handwashing point.** There should be one handwashing point per 40 students in schools and per 30 students in dormitories; for kindergarten schools, there should be one handwashing point per 20 children.

(iii) **Toilet for girls.** There should be one toilet for every 30 girls in schools, with at least one separate toilet provided for female staff. For kindergartens, there should be one toilet for every 15 children.

(iv) **Toilet and urinal for boys.** There should be at least one separate toilet and urinal per 40 boys, and at least one separate toilet and urinal for male staff.

3. Water Quality

Potable water (water for drinking or use in food preparation and cooking, and/or personal hygiene) should meet the Mongolian water quality standards. Water required only for nonpotable uses, including cleaning, laundry, and sanitation, does not have to meet these standards.

The key standards include the following:

(i)   No *Escherichia coli* (*E. coli*) or thermotolerant coliform bacteria should be detectable in any 100 milliliter sample.

(ii)   Chemical and radiological parameters should not exceed the following values: arsenic concentration in drinking water should not exceed 0.01 milligram per liter (mg/l) (10 micrograms per liter [µg/L]); fluoride concentration in drinking water should not exceed 1.5 mg/l; and total dissolved solids (TDS) should preferably be less than 1,500 mg/l.

(iii) Hardness should preferably be less than 200 mg/l—equivalent to about 4.25 milligram-equivalents per liter (mg-eq/liter) on the Russian scale commonly used in Mongolia, and 12 degrees of hardness (dGH) on the German scale commonly used in Western Europe. It may be necessary to accept higher levels where groundwater is hard and there is no alternative water source.[2] There should be no taste, odor, or color that would discourage people from consuming the water.

4. Quantity of Water Available

Water is required for drinking, cooking, personal hygiene, laundry, and toilet flushing (where water-flushed toilets are provided). The normal measure of water to be provided is the volume per person (capita) per day, given as liter per capita per day (LPCD). Table 1 shows guideline figures for minimum water availability.

Table 1: Guideline Figures for Minimum Basic Water Availability in Schools in Mongolia

Minimum Per Capita Water Requirement	For Regular Students (LPCD)	For Students Living in Dormitories (LPCD)
Basic requirement	7–8	22
Additional for conventional flush toilets	10–15	10–15
Total for waterless toilets (direct-drop and ecological designs)	7–8	22
Total for conventional flush toilets	17–25	32–37

LPCD = liter per capita per day.
Source: Ministry of Education, Culture, Science and Sports, Ministry of Health, and Ministry of Finance (2015).

The basic requirement for regular students covers water used for drinking and hand hygiene. In addition, students living in dormitories need water for washing, food preparation, and laundry. Water should be available throughout the school day, which means the whole day for students living in dormitories. To meet the requirement, storage will normally be required.

---

[2]   Hardness in water is not in itself harmful but can make water unpalatable when present at very high concentrations. Water in Southern Mongolia typically has a hardness of around 7 mg-eq/liter, which falls into the "hard" category. Regardless of theoretical measurements, the best measure of whether or not water is drinkable would be the views of local people, including schoolchildren.

For regular students (i.e., students who commute daily to schools), the World Health Organization (WHO) also sets similar minimum water availability standards (Adams et al. 2009). For students living in dormitories, WHO recommends doubling the additional allowance for conventional flush toilets. In the case of using pour-flush toilets, WHO suggests an additional 1.5–3.0 liters per regular student per day, and 3.0–6.0 liters per student living in a dormitory per day. The total water requirement for poor-flush toilets would then be 8.5–11.0 liters per regular student per day and 25.0-28.0 liters per student living in a dormitory per day.

5. Provision for Small Children and the Disabled

It is important that toilets and handwashing points in schools cater to the needs of small children. In particular, taps, handwashing points, and soap should be located at a level that small children can reach. Similarly, it is essential that handwashing and toilet facilities in schools are designed to take account of the needs of disabled children, as children with disabilities are often discouraged from attending school if they find that WASH facilities are inaccessible (Erhard et al. 2013; Emory University and UNICEF n.d.). According to the minimum requirements, all schools, kindergartens, and dormitories should have at least one latrine that is suitable for students with disabilities. Further, facilities should be provided with handrails and ramps.

6. Provision for Adolescent Girls

The minimum requirements for WASH specify that there should be one washing and/or changing room for girls. To ensure privacy, the room should have a lockable door and clean water supply, and contain a waste bin for disposal of sanitary pads, which should be emptied daily.

The lack of suitable facilities for girls is likely to have significant negative effects on education and health outcomes. Studies by Emory University in the Kyrgyz Republic and Uzbekistan found that girls were affected by the poor maintenance of facilities and lack of privacy, rather than by overall lack of basic access (Emory University and UNICEF, n.d.). The lack of toilet doors and proximity to boys' latrines discouraged girls from using school WASH facilities, with potential deleterious health effects, including dehydration and physical discomfort. Lack of privacy and the absence of adequate access to suitable facilities are particularly problematic for menstruating girls and could lead to higher absenteeism and the possibility of higher drop-out rates.

7. Hygiene Education

The minimum requirements also cover hygiene education, as follows:

(i)   Hygiene education, covering handwashing, tooth brushing, waste management, and proper usage of WASH facilities, should be part of the school curriculum.

(ii)   Hygiene education materials should be available and appropriate to the age of students and their development needs.

(iii) Schools should play a role in raising public awareness on health-related matters.

(iv)   Each school should have a fully trained staff member, designated as the focal point for WASH-related issues.

(v)   Schools should support the establishment of clubs to raise awareness of WASH issues among students and encourage better hygiene practice.

(vi)   An internal monitoring system should be established to assess changes in the attitudes and behavior of students after improvements in WASH facilities and students' exposure to hygiene education initiatives.

C. System Needs

Fulfilling these requirements is dependent on the existence of good management systems at the school level and effective support from the government. The lack of physical, institutional, and financial resources in many rural locations often makes it difficult and costly to raise WASH facilities and services to the required standards.

For example, water-flushed toilets require a reliable source of water close to the toilets. If a school does not have a piped water supply, the water-flushed toilets installed at the school will fail because of the lack of water. In a similar vein, a school water supply system may become nonoperational if the design of WASH facilities focuses only on water outlets, in the form of taps and connections to toilet cisterns, without attention to the systems required to deliver water to those outlets.

To meet the requirements, it is therefore important to look beyond taps and toilets and consider complete water supply and sanitation system needs for delivering water to users, and removing and safely disposing of, or reusing, excreta and wastewater.

There is a need to develop capacity at the *aimag* level to gather and organize information, identify priorities, and support WASH planning exercises. *Aimag*-level actions should include effective coordination among the departments that must be involved in planning and subsequent management of WASH facilities and services. There is also a need to develop systems to provide support to schools and organize tasks, such as periodic desludging of septic tanks, that are beyond the capacity of individual schools.

III. Planning for Improved Water, Sanitation, and Hygiene in Schools

The objectives of improving water, sanitation, and hygiene (WASH) in schools are for students, teachers, and school staff to have access to safe water supply and sanitation facilities and services that meet at least the minimum requirements presented in section II; and to use those facilities and services in ways that enhance their health and well-being.

Lasting improvements in health and well-being are only possible if facilities and services continue to meet these objectives over time.

The overall process for requesting funding and support from the government to improve WASH in schools in Mongolia involves the following steps:

(i)   The school identifies its WASH needs and consults the *aimag* or Ulaanbaatar education department on possible sources of funding and support. The main sources of funding are the Local Development Fund of the local government, or the budget of the central government.

(ii) The *aimag* or Ulaanbaatar education department considers all such needs, assesses priorities, and prepares a prioritized list of schools with WASH improvement needs. Depending on the funding source, the *aimag* or Ulaanbaatar education department submits the list to the local government (Local Development Fund) or the Ministry of Education, Culture, Science and Sports (MECSS, central government's budget). The request is further reviewed and prioritized by the local government, MECSS, and/or the Ministry of Finance for funding.

(iii) Once the request has been approved for funding, the local government or MECSS launches the bidding process for engineering design, drawings, and cost estimates. These designs, drawings, and cost estimates need to be approved by relevant government agencies before the bidding process for civil works is launched.

Based on the overall process described here, this section guides various stakeholders in planning improved WASH in schools.

A. Assessing Needs and Options for Improved Water, Sanitation, and Hygiene in Schools

Planning for WASH services should start from assessment of the need for WASH facilities and services, stated in terms of the number of facilities required and the location of those facilities within the school. Toilet, washbasin, and shower requirements depend on the number of regular students, students staying in dormitories, teachers and staff at the school, as well as the design standards set out in section II. Toilets, showers, and washbasins located in sanitation blocks should be accessible from school buildings, including dormitories. Existing facilities should be incorporated into the WASH improvement plan if they are in good condition and compatible with the planned WASH improvements.

The use of standard designs saves time and ensures a degree of uniformity. However, standard designs are less appropriate for WASH facilities, particularly for schools in remote rural areas, for the following reasons:

(i)   Many schools are located in areas with poor water resources. Following standard designs without considering water availability may result in a situation where pipes are laid but there is no water to supply them.

(ii)   Standard designs are likely to assume conditions found in Ulaanbaatar and *aimag* centers and depend on the existence of district heating systems for their success, which are rarely available in rural *soums* and *baghs,* and in recently developed peri-urban settlements.

It is therefore important to assess if the school has access to water supply and sanitation systems, as these determine the level of service that is possible. Factors that affect the level of service for water supply include the volume and quality of the water that is available to users, the reliability of the water delivery system, and the locations at which water is available. Where water availability is limited, as is the case in some rural areas in Mongolia, it may be necessary to accept a lower level of service than would be the norm where water is readily available.

For sanitation, an important determinant of level of service is the location of the toilet. Toilets located inside buildings are more convenient than outside toilets and can offer a higher level of service. Water-flushed toilets incorporating a water seal provide a higher level of service than dry direct-drop toilets, because a functioning water seal will greatly reduce nuisance from odors and insects. However, where water availability is limited, a flush toilet may cease to function well and deliver a lower level of service than a well-managed dry system.

Options for disposal of wastewater, including black water from toilets and gray water from kitchens and bathrooms, will be influenced by the volume of water delivered and the type of sanitation facilities provided. In terms of level of service, the basic requirement is that all wastewater should be removed from the local environment and disposed of in ways that minimize contact with people. Where water availability is limited, the options for reusing gray water for purposes such as toilet flushing may need to be considered.

For solid waste disposal, the basic requirement is that all waste is removed from the local environment and disposed of safely. Factors that affect the level of service include the frequency of collection and the location from which public collection services lift waste.

The school located in an area with public water supply and sewer systems should connect its WASH facilities to those systems. The focus of planning activities can then be on the school's internal WASH systems.

Where there is no public water supply, but the school is close to a reliable and good quality water source, the preferred option will normally be to develop a scheme to serve the school based on the available water source. Provided that it has an acceptable mineral content, groundwater will be the preferred source, since unlike surface water and spring water, it is unlikely to be heavily contaminated with pathogens.

If the nearest viable water source is some distance from the school, the best option will usually be to prepare a scheme to provide water to the settlement in which the school is located. The school will not normally play the lead role in planning and managing the settlement-wide scheme.

With a potential water source identified, the focus of planning must turn to systems for extracting water from the source and delivering it to the school. Where possible, pumps should be powered by electricity from the national grid. For schools in remote *baghs* and *soums* that do not have access to the electricity network, solar and wind power should be investigated as alternatives to expensive diesel.

Where connection to a public sewer is not possible, some form of on-site storage and treatment will be required. Complex mechanical treatment systems serving individual schools should only be considered if arrangements can be implemented to ensure their effective operation and maintenance over time.

Because of Mongolia's harsh winter, transmission mains, distribution mains, and storage tanks must be protected against freezing. Burying pipes at depths alongside an existing or planned district heating system should thus be considered where possible, as this option has relatively low operational costs despite its high initial capital costs. Storage tanks should also be well insulated and located within buildings so that the heat from the building will help to prevent freezing. As with water supply, sewers, septic tanks and treatment facilities must not freeze during the winter.

Lastly, it is important to ensure that people use WASH facilities correctly. For example, handwashing facilities should be available close to toilet facilities so that people can wash their hands after defecating. Without handwashing facilities close to toilets, it will be difficult for people to maintain good hygiene practices. Although the requirement to locate handwashing facilities close to toilets is included in the Mongolian construction code, it is often neglected. This is particularly true when sanitation takes the form of dry pit-latrine toilets, situated some distance from school buildings, which is usually the case at schools in rural areas. Without good hygiene, the expected health benefits from water supply and sanitation improvements will not materialize. Therefore, hygiene education should always be a component of WASH plans. Hygiene education should focus on a small number of key messages, including the need for handwashing with soap.

Water supply, sanitation, and hygiene education options should be assessed in relation to systems for ensuring that WASH facilities continue to deliver good quality services. Where such systems are currently not in place, actions will be needed to strengthen and improve institutional systems. Once options have been assessed and preferred options have been identified, they should be incorporated into a WASH plan for the school, which would then provide the basis for detailed design and costing.

Figure 1 sets out the steps in planning, improving, and managing WASH facilities, including hygiene education. It also indicates which sections of this Guide provide detailed information on the steps.

Figure 1: Steps in Planning, Improving, and Managing Water, Sanitation, and Hygiene Facilities

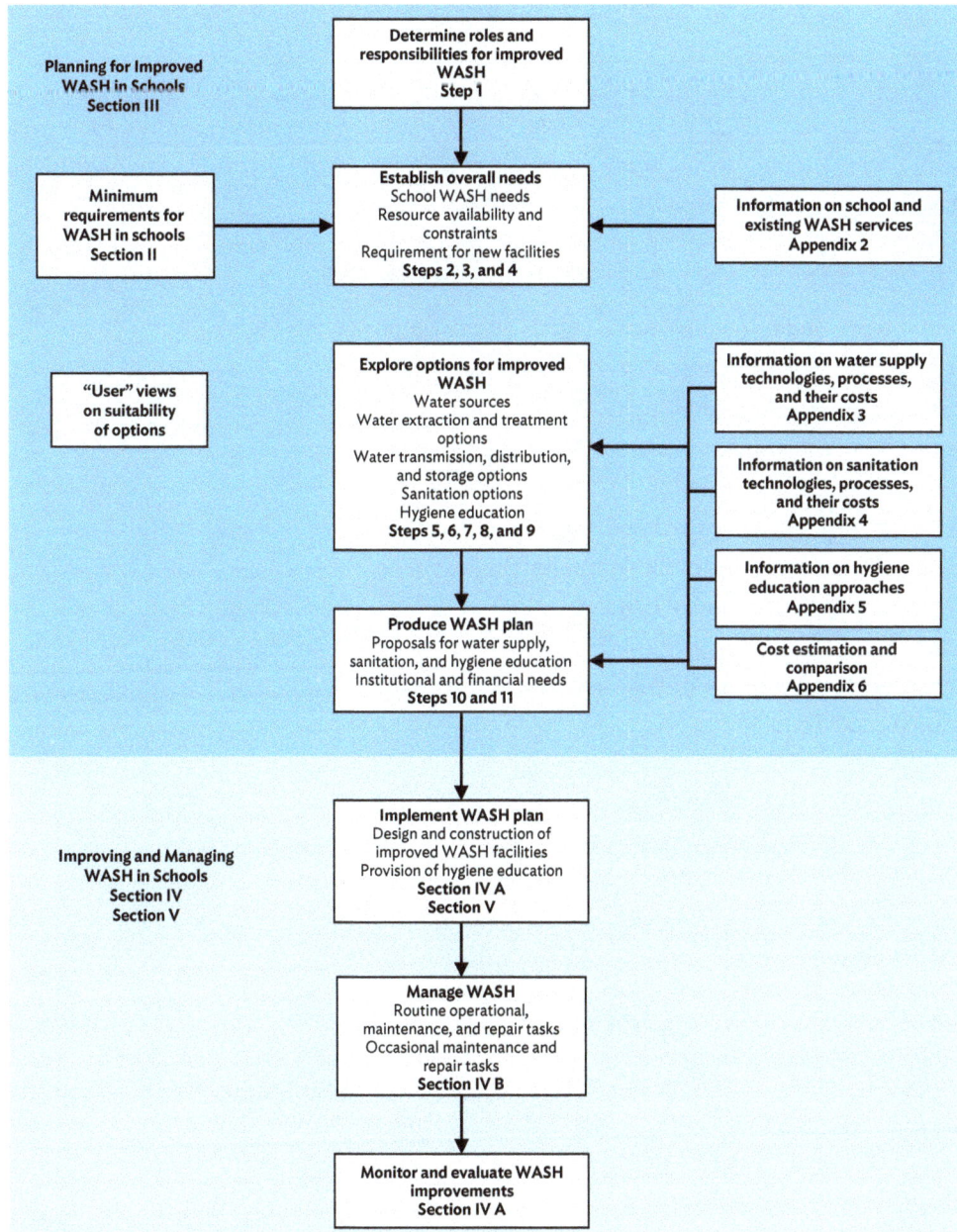

WASH = water, sanitation, and hygiene.
Source: Authors.

B. Steps in Planning Improved Water, Sanitation, and Hygiene in Schools

The WASH plan should adopt an integrated approach to physical improvements and hygiene education and identify any institutional improvement required to overcome constraints that might impact on successful delivery of WASH services. Without such an integrated approach, there is no guarantee that the provision of improved WASH facilities will result in improvements in health and well-being.

The process of planning for improved WASH involves the following steps:

(i)   determine roles and responsibilities of stakeholders in improved WASH;
(ii)   establish school WASH needs;
(iii)   identify resources and constraints;
(iv)   determine the requirements for new WASH facilities;
(v)   assess water sources;
(vi)   assess water extraction and treatment options;
(vii)   assess water transmission, distribution, and storage options;
(viii) select an appropriate sanitation option;
(ix)   design hygiene education activities;
(x)   assess institutional and financial needs; and
(xi)   prepare the WASH plan.

Step 1: Determining Roles and Responsibilities of Stakeholders in Improved Water, Sanitation, and Hygiene

Various stakeholders in school WASH services should be involved in the planning process at appropriate times and in appropriate ways.

It is essential that the intended users of services, students, teachers, and staff, as well as school management, are consulted at the planning stage, not so much on technical aspects of WASH as on their views on WASH-related priorities and subjects, such as the location of WASH facilities, which will affect their use. Particular attention should be paid to the needs of small children and students with disabilities who should be consulted to ensure that proposed WASH facilities are appropriate to their needs.

The extent to which various stakeholders should be involved will depend on the existence of reliable public water supply. Reliable public water supply, if available, is normally the preferred option for supplying the school. The focus of planning efforts will then be on system improvements within the school boundary. The public provider of water supply will be involved in the planning process but only to confirm that they can provide water in the quantity required by the school.

Where a public water supply does not exist, as is the case with many schools in rural areas, the options will be to develop a scheme that serves only the school, or serves the whole settlement (*soum, bagh, ger* area) within which the school is located.

The first option will be possible if there is a viable water source close to the school. The school choosing this option will be responsible for making its own WASH arrangements, engaging engineers to plan and design the scheme, as well as management.

The second approach will be more appropriate if a water source is located some distance from the school and settlement. Water supply to the school will then depend on actions to address the need for water supply in the *soum, bagh,* or *ger* area where the school is located. The range of stakeholders to be involved in the whole settlement approach will be wider than that in the school-only approach. Assessment of water availability will usually require a detailed survey of potential water sources in the form of a feasibility study.

Step 2: Establishing School Water, Sanitation, and Hygiene Needs

The assessment of WASH needs should be guided by the minimum requirements for WASH in schools jointly issued by MECSS, Ministry of Finance, and Ministry of Health in 2015 (section II) and should cover existing facilities.

Overall, WASH needs at the school should be assessed in terms of

(i)   type of WASH facilities required,
(ii)   locations where WASH facilities are required;
(iii)   number of WASH facilities required at each location; and
(iv)   quantity and quality of water required.

Further guidance on information collection and analysis for assessing WASH needs is in Appendix 2.

Where existing facilities are in poor condition and of low quality, new facilities will normally be required. If existing facilities—washbasins, shower units, and toilets—are in reasonably good condition and of adequate quality, the potential for using these facilities in the new scheme should be assessed, even when some of them are currently nonoperational. Existing facilities that are usable or can be rehabilitated could be included in the scheme, reducing the need for new facilities.

a. Type of Water, Sanitation, and Hygiene Facilities Required

WASH facilities, including washbasins, shower facilities, and toilets, should be grouped in sanitation blocks. Facilities for boys and girls should have separate entrances and be separated by solid walls that reach up to the roof. The doors of individual toilet cubicles should be securable and reach down to the floor. Appropriate provisions should be made for small children, students with disabilities, and adolescent girls.

Specifically, required WASH facilities include the following:

(i)   washbasins and shower facilities to serve both regular students and students living in dormitories;
(ii)   water outlets to provide water for flushed toilets and handwashing for all types of toilets;
(iii)   water outlets at kitchens, to allow handwashing and provide water for use in food preparation; and
(iv)   toilets, either dry or water-flushed, depending on water availability (both urinals and toilets should be provided for male students and teachers).

b. Locations Where Water, Sanitation, and Hygiene Facilities Are Required

There should be one centrally located block of toilets and washbasins to serve regular students. The block should be located close to classrooms and the area where students eat, so that students can wash their hands before eating.

To serve dormitories, sanitation blocks providing toilets, washbasins, and showers will be required. The preferred arrangement will be for each dormitory building to have its own attached block, while the possibility of providing one sanitation block to serve two or more closely spaced dormitories could be considered. Where a sanitation block is not directly accessible from the dormitories that it serves, enclosed and lighted access routes between dormitories and the sanitation block will be required to facilitate student access to the block under all weather conditions.

When assessing possible locations for sanitation blocks, the location of all classrooms, dormitories, and other buildings should be considered, together with features such as steep slopes and access routes that would influence the location of facilities. It is important to consult students, teachers, and other intended users about possible locations, and to take their views into account when deciding preferred locations. One approach could be to encourage teachers and students to prepare a sketch showing the location of school buildings, existing water supply and sanitation facilities, and potential sites for future facilities. Such a sketch could then be used as a basis for discussion on possible locations of planned facilities. The results of this exercise can be refined with the help of engineers.

Water outlets and arrangements for draining waste gray water will be required at kitchens and laundry facilities. The facilities at kitchens should include one or more sinks for washing vegetables and cleaning utensils, and separate washbasins where kitchen staff can wash their hands. If possible, separate toilet facilities should be provided for kitchen staff. Provision should be made for heating the water used at kitchens and laundries.

c. Number of Water, Sanitation, and Hygiene Facilities Required at Each Location

The total number of washbasins, shower facilities and toilets required will depend on minimum requirements for WASH in schools as issued jointly by MECSS, Ministry of Health, and Ministry of Finance in 2015 (see section II); and the number of regular students and students living in dormitories, and teachers and staff at the school.

The procedure for calculating the total number of units required is as follows:

(i) Obtain information on the number of regular students, teachers, and staff at the school; the number of students living in dormitories at the school; and the number of small children (6–7-year-olds).

(ii) Assess the likely increase in the number of students, teachers, and staff for the lifespan of the planned WASH facilities (typically 20 years), taking account of the possibility that improved WASH facilities will result in increased interest among herder families to send their children to the school.[3]

(iii) Calculate the number of toilets and washbasins required by dividing the total number of students by the permissible number of students per unit as set out in the minimum requirements for WASH in schools

---

[3]   An evaluation report on the school WASH project supported by the UNICEF and AusAID in Khuvsgul *aimag* referred to *soum* leaders and school principals who reported that herder families are willing to send their children to the schools where WASH facilities had been improved under the project.

(see section II). Consider additional toilets for male and female staff and appropriate provisions for small children and students with disabilities in accordance with the minimum requirements.

(iv) Separately calculate the number of toilets, washbasins, and showers required to serve individual dormitories based on the number of beds in each dormitory and in accordance with the minimum requirements. For toilets and washbasins intended for small children, the height should be adjusted to suit their needs.

(v) Calculate the number of facilities required for male and female students separately. At least one toilet for students with disabilities should be provided in each sanitation block.

d. Quantity and Quality of Water Required

The average daily water requirement for the whole school can be calculated using the estimated number of students and teachers, and the minimum basic water availability per student, as previously shown in section II (Table 1). When assessing system requirements, allowance must be made for fluctuations in demand over the day and during the year. A peak factor should be applied to make allowance for fluctuations in demand.[4] Water requirements should be divided between sanitation blocks, depending on the type of facilities and number of students using each block.

To ensure that the water delivered to taps is safe to drink, the quality of water should meet Mongolian standards (section II). This will normally require treatment.

Further guidance on calculating demands and assessing water treatment options is found in Appendix 3.

Step 3: Identifying Resources and Constraints

Lack of resources constrains successful implementation of planned WASH improvements. When assessing the options for providing water supply and sanitation systems, it is important to examine resource availability.

Resources to be considered include the following:

(i) **Water.** Water is the most important physical resource, and a constraint, if no easily accessible water source exists.

(ii) **Energy.** Energy is required to power pumps, heat pipes, and operate water and wastewater treatment plants. Electricity is normally the cheapest form of energy for operating pumps and other equipment, and lack of a networked electricity supply may be a constraint. Most public heating systems in Mongolia use either solid fuel or oil to heat water pipes.

(iii) **Technical knowledge and skills.** These are required to operate WASH systems, particularly those that are more technically complex.

(iv) **Management knowledge and skills.** These are required to ensure continuously effective delivery of WASH services and prompt action to deal with any problem and changes in circumstances that may arise.

(v) **Knowledge of hygiene and skills.** These are required to design and carry out hygiene education activities.

---

[4] If no local information is available, a peak factor of 1.5 may be applied when calculating the water production capacity needed (Appendix 3 provides further guidance).

(vi) **Funds.** Funds are needed for the operation, maintenance, and repair of WASH facilities and the conduct of hygiene education activities. The main source of funds is likely to be the government budget, though other sources may be available.[5]

The most important constraint is the lack of a viable water source close to the school. However, even if an apparently accessible water source exists near the school, other constraints could also impact on water supply and sanitation services (Box 1).

Box 1: School Water Supply in Khatgal Bagh, Khuvsgul Aimag

Khatgal is a *bagh* of Alag-Erdene Soum in Khuvsgul Aimag, northern Mongolia. It had no functioning centralized water supply system in 2011. Like other rural settlements in Mongolia, Khatgal had had a centralized water supply system before the country started to transition from a centrally planned economy to a market economy in 1989. The system in the pretransition era fell into disrepair during the initial transition years and had become no longer operational.

In 2011, the school in Khatgal Bagh had about 600 students, with 120 living in dormitories. The school is situated less than 2 kilometers from the Khuvsgul Lake, which contains 1.5% to 2% of the earth's fresh water. Water was delivered to the school by horse-drawn cart each day. Because the amount of water delivered to the school was small, there was only about 1.7 liters of water available per student per day.

Despite its proximity to the lake, the school still lacked an adequate water supply. Though its sanitation facilities were improved, the school had only dry "VIP (ventilated improved pit)" toilets in 2015.

Source: AusAID (2011).

Resource constraints on water supply and sanitation services also include

(i)   lack of a dedicated budget to cover the ongoing operation, maintenance, and repair of WASH facilities;

(ii) limited availability of the technical and management knowledge and skills required to ensure sustainable operation of WASH facilities; and

(iii) lack of a reliable electricity supply.

The plan for improved WASH should include actions to address constraints identified in the assessment. If existing WASH facilities have failed, it is important to examine the reasons for the failure.

---

[5]   Other sources include official development assistance, local development funds, private sector, and international nongovernment organizations (NGOs). Examples include an initiative by the international NGO World Vision, the telecommunications company Mobicom, and *aimag* governments to provide improved sanitation facilities in a number of schools. According to the World Vision's report, the initiative resulted in sanitation improvements benefiting at least 15,000 children. *World Vision International. Improved Sanitation—No More Latrine!* https://www.wvi.org/mongolia/video/improved-sanitation-no-more-latrine.

Step 4: Determining the Requirement for New Facilities

School WASH facilities are only a part of overall water supply and sanitation systems. If the other components of the systems are either absent or inadequate, WASH facilities at the school will remain unused, which is not uncommon in rural schools. It is therefore important to consider complete water supply and sanitation systems at the planning stage.

a. Water Supply

The most important factor to consider when assessing water supply options is the water source. Without a conveniently located source that can meet the school's needs for water supply at an affordable cost, it will not be possible to supply enough water to meet even the minimum requirements and satisfy standards. Affordability is a critical factor because systems that are expensive to operate due to the distance to the source, or the need for treatment, will be difficult to sustain. Similarly, technologies that require operational knowledge and skills that are not locally available will quickly fail unless actions are taken to provide appropriate operational support systems.

The system to deliver water from the source or sources to users would include some or all of the following, depending on the type and location of the source:

(i) **An intake, springbox, or well.** This is provided to allow water to be extracted from the source.
(ii) **Provision for treatment.** This will be required if water in the source does not meet the drinking water quality standards.
(iii) **Pumps.** These will be required whenever it is not possible to deliver water from the source to the school by gravity.
(iv) **A transmission main.** This will be required where the source is located some distance away from the school.
(v) **Storage tanks.** Water can be distributed to users from these by gravity.
(vi) **Distribution pipes.** These will be used to deliver water from reservoirs to water outlets.
(vii) **Water outlets**. These will be mainly taps but will include connections to cisterns if sanitation facilities include cistern-flushed toilets.

Wherever possible, the school should be supplied from a public water supply system.

b. Sanitation

Although safe sanitation systems start from well-designed, appropriate toilet facilities, provision must be included for the removal, transport, treatment, and reuse or disposal of excreta. The need for complete sanitation systems is best approached through the concept of the sanitation chain that includes some or all of the following:

Capture–Storage–Removal–Transport–Treatment–Reuse or Safe Disposal

(i) **Capture.** This is achieved through the toilet. For water-flushed systems, the toilet usually takes the form of a squat or seat-type water closet incorporating a water seal. For dry systems, the toilet always involves a simple direct-drop arrangement, which could also incorporate provision for urine separation.
(ii) **Storage.** This is required for all nonsewered systems, including dry systems and water-flushed systems that retain solids on-site. In the case of pit latrines, leach pits, and septic tanks, solids are normally stored for a period of at least a year. Liquid is allowed to seep into the ground or, in the case of urine-diversion

toilets, is collected separately. Containerized dry systems store excreta for much shorter periods, typically less than a week. The container volume required will be reduced if urine can be diverted and dealt with separately.

(iii) **Removal and transport.** Arrangements for these depend on whether solids are retained on-site. On-site and containerized systems require periodic removal of fecal sludge, which would then be transported to a suitable treatment site, using a vacuum tanker or other suitable form of transport. Almost all off-site systems use sewerage to transport both solids and liquids from toilets, sinks, and baths away from school and residential areas. Sewers only work if connections exist to carry wastes from water-using appliances to the sewers.

(iv) **Treatment.** The need for treatment depends on the strength of the wastewater to be treated. Fecal sludge and septage (combination of fecal sludge and supernatant water removed from septic tanks and wet pits) are much stronger than the sewage produced by schools and households.

More detailed information on each link in the sanitation chain is available in Appendix 4.

Step 5: Assessing Water Sources

Once the needs have been established in terms of number of sanitation blocks and other water-using facilities required, attention should turn to assessing options for supplying the water required for those facilities to function.

The assessment should seek answers to the following questions:

(i) From where does the school obtain water at present? What information is available on the adequacy of this source in terms of both quantity and quality?

(ii) Is piped water available near the school? Does the piped water system have enough capacity to support an improved supply to the school?

(iii) What other potential sources exist? How far are they from the school? What information is available on the likely yield and quality of water from each potential source? What capital and recurrent costs would be incurred to transport water from the source to the school?

Where there is an existing piped water supply near the school, it is usually best to supply the school from the piped water system. However, many rural schools do not have access to a piped water supply. If that is the case, other possible sources must be identified and assessed. If there is more than one possible source, each possible source should be assessed with reference to:

(i) water yield,
(ii) cost of delivering water from the source to the school, and
(iii) complexity of the management arrangements required to extract and convey water from the source to the school.

Figure 2 shows a logical procedure for carrying out the assessment. It starts from the premise that examination of options should start with those that are technically and institutionally simple. More complex options should be considered only if local conditions do not allow adoption of a simpler option. More detailed guidance for assessing water supply options is provided in Appendix 3.

Figure 2: Procedure for Assessing Water Supply Options

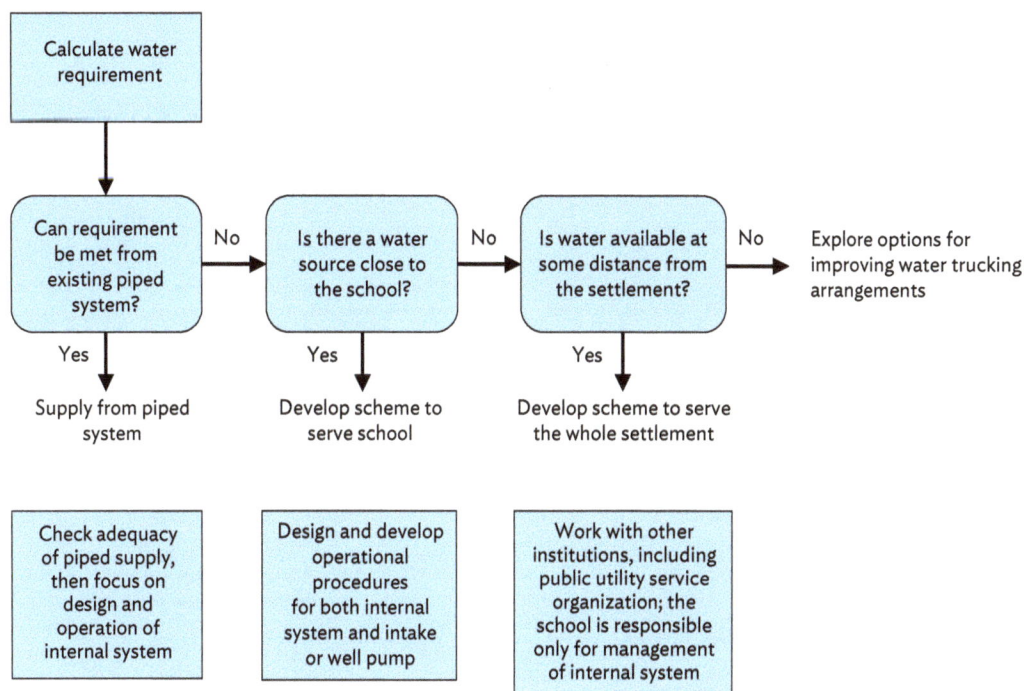

Source: Authors.

When the water source is some distance away from the settlement where the school is located, it will be expensive to implement a scheme that only serves the school. Under such circumstances, the best option will be to develop a scheme to serve the whole settlement. This option will require cooperation between government agencies and service providers and will depend on the availability of funds beyond those available for education administration. Securing the required funding and reaching agreement on roles and responsibilities for implementation and subsequent operation of the new water supply scheme will often take time.

If a settlement-wide scheme cannot be implemented in the immediate future, the strategy should be to identify and implement improvements that can be made immediately, using the existing water source and working with other stakeholders to develop an improved water supply scheme that will serve the whole settlement in the long run.

Step 6: Assessing Water Extraction and Treatment Options

Once one or more potentially viable water sources have been identified, options for extracting water from each source and delivering it to water outlets at the school should be assessed. If the quality of the source water does not meet the Mongolian standards, treatment will be required.

a. Water Extraction

The arrangements for extracting water from a source will depend on the type of source and be influenced by the availability of electricity and technical knowledge and skills to operate and maintain the arrangements. Groundwater usually requires pumping.

Table 2 summarizes options for pumping groundwater, with information on advantages and disadvantages for each option.

Table 2: Summary of Options for Pumping Groundwater

Pumping Option	Advantages	Disadvantages
Hand pumps	• Cheap and not dependent on power • Easy to maintain	• Limited capacity, insufficient for piped distribution system • Only possible if groundwater is available below the school
Diesel-powered pumps	• Cheap to purchase • Maintenance and repair are easy	• High operating costs • Dependent on reliable supply systems for diesel that may be difficult to guarantee in rural remote areas
Electrically powered pumps with electricity supplied from grid	• Cheap to purchase • Operational costs are lower than those of diesel-powered pumps	• Dependent on reliable 24-7 electricity supply that may not be available in some locations
Solar-powered pumps	• Not dependent on remote power source or purchase and transport of diesel • Operating costs are low	• High initial capital costs • Limited capacity at present • Diesel back-up may be needed for periods when solar radiation is limited
Wind-powered pumps	• Not dependent on remote power source or purchase and transport of diesel • Operating costs are low	• High initial capital costs • Mechanically complex • Diesel back-up systems may be needed for maintenance and repair

Source: Authors.

Electrically powered pumps will be the best option if the school is connected to the electricity grid. Because of their limited capacity, hand pumps should be considered only as a temporary measure to improve an existing well where there is no piped water system. Hand pumps should be replaced by a better system at the earliest.

Most schools in *soum* centers and *baghs* without connection to a centralized electricity network rely on diesel-powered pumps. These pumps are expensive to operate, particularly when diesel has to be transported over long distances. Under such circumstance, the most effective way of reducing operational costs would be to extend the electricity network to the settlement where the school is located and to power pumps with electricity delivered from the grid. If this option is not possible, solar and wind power options should be explored. Although both have high initial capital costs, these costs are reducing rapidly and their operational costs will be lower.[6]

Deep groundwater requires treatment only if it is hard or has a high mineral content. If groundwater is unpalatable because of high hardness or salinity, it is better to investigate other water source options before deciding on a scheme that requires treatment to remove minerals. Disinfection is desirable to remove any risk of contamination and pathogen transmission in storage tanks and distribution systems.

Some schools may be close to springs and surface water. Extraction from a surface water source requires an intake structure, which should be designed in a way that it continues to operate when the source is frozen. That is, the intake should be designed to take water from a level below the bottom level of the ice layer that forms over the source during the winter.

---

[6]   Further research is needed to determine whether extreme cold conditions affect their viability in any way.

Surface water is usually contaminated and requires treatment before use in a potable water system. The photo below shows a polluted spring that requires protection and treatment to improve water quality. Water from all spring and surface water sources should be disinfected before delivery to users. One possible exception would be mountain streams that rely on perennial springs or melted snow or ice, and can be accessed above the point at which human and animal pollution is possible.

Polluted spring. *A spring near Bayan-Uul, Govi-Altai Aimag is a polluted water source that requires protection and treatment to improve water quality* (photo by Kevin Tayler).

Water from springs and surface water sources usually requires pumping after extraction, with pumps located close to the intake or spring. The pumping options in Table 2 also apply to springs and surface water.

b. Water Treatment (Including Disinfection)

Treatment is required if the quality of the source water does not meet Mongolian standards (section II). Deep groundwater is unlikely to contain pathogens and will normally be the first choice if it is uncontaminated by either chemicals or minerals. All other sources need treatment, varying from disinfection to full treatment including sedimentation and filtration followed by disinfection. Figure 3 presents a guide for assessing water quality and treatment requirements.

Figure 3: Assessing Water Quality and Treatment Requirements

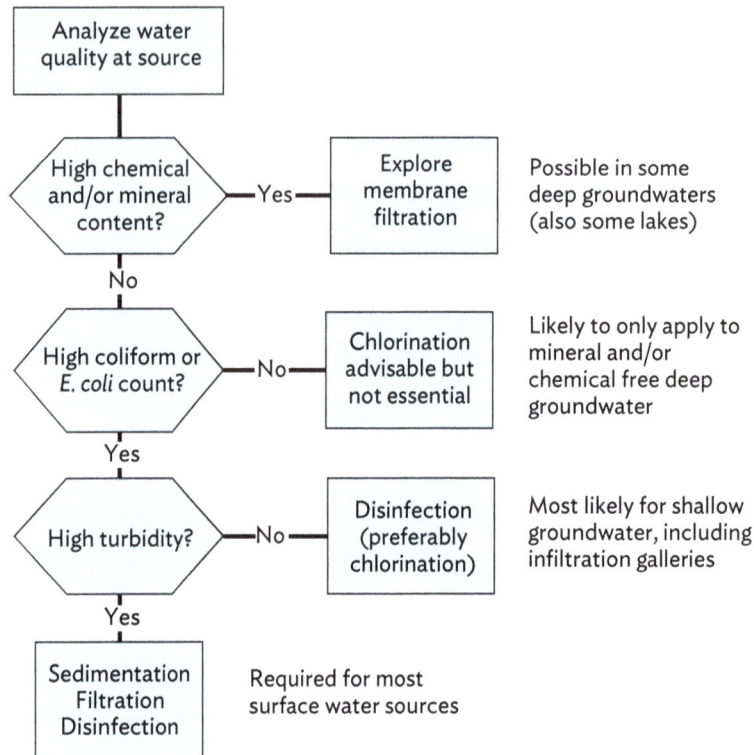

Source: Authors.

Treatment can be provided prior to distribution or at the point of delivery.

(i)   **Treatment prior to distribution.** This is usually the cheapest option but will be difficult to sustain if the technical and management skills required to operate a treatment plant installed at the school or settlement level are lacking.

(ii) **Treatment at the point of delivery.** This could be more expensive but devolves operational responsibility to those who use a kitchen or washroom. The link between responsibility and use may make it easier to ensure good system management.

For schemes that are intended to meet only the school's needs, treatment at the point of delivery is likely to be more viable than treatment prior to distribution. For schemes that serve the whole settlement, treatment is normally required prior to distribution and is the responsibility of government agencies rather than the school. The design of treatment facilities to serve the whole settlement requires detailed technical knowledge and is not considered further in this publication.

Appendix 3 provides more detailed guidance on assessing water extraction and treatment options and developing schemes based on preferred options.

Step 7: Assessing Water Transmission, Distribution, and Storage Options

Water transmission is required only if the source is remote from the school. Regardless of whether a transmission main is required or not, key points to consider when assessing water system options are

(i)   main sizes and materials,

(ii)   options for preventing failure due to freezing in the system, and

(iii)   storage requirements in terms of location and volume.

a. Main Sizes and Materials

Main diameters depend on calculated design flows, which in turn depend on per capita water demand, number of people served, configuration of the distribution system, and relationship of the average flow to the peak flow through the main. No main serving a whole sanitation block or kitchen area should be less than 50 millimeters in diameter. Internal mains serving washbasins, sinks, and toilet cisterns can have smaller diameters.

Materials commonly used for water distribution pipes include ductile iron, uPVC, high and medium density polyethylene, and cross-linked polyethylene (PEX). Copper is the most commonly used material for internal pipework. Ductile iron, which is expensive, is only an option for larger diameter pipes, unlikely to be used for Mongolian school water supply systems (Appendix 3).

b. Options to Prevent Freezing

Flexible pipe materials such as polyethylene are preferable to rigid materials such as uPVC because they can accommodate some expansion when water inside the pipe freezes. If their contents are subjected to several freeze–thaw cycles, however, even flexible pipes are likely to fail.

Below are options for preventing freezing with their advantages, disadvantages, and limitations (more information is available in Appendix 3).

(i)   **Bury pipes below the depth at which the ground freezes.** The construction cost of this option will be high, while operational costs will be low. One potential problem will be the cost and difficulty associated with repairing pipes. Using long lengths of polyethylene pipe will reduce the potential for leaks and the need for pipe repairs. This option is not possible in areas that are subject to permafrost.

(ii)   **Heat pipes.** Pipes can be heated by using an electrically powered heat trace or laying pipes inside or alongside a heated conduit (known as a utilidor). The heat trace option will have a low capital cost but will be expensive to run. The utilidor option is only viable in those areas of Ulaanbaatar, *aimag* and *soum* centers where district heating is provided or planned, and in schools that have their own centralized heating systems.

(iii)   **Circulate heated water through a looped system.** This approach requires the circulated water to pass through a heat exchanger. Its drawbacks are its relative complexity and its high operating cost.

c. Storage

Storage is required to allow for variations in water demand. It fulfills the following functions:

(i)   Balance variations in supply and demand with storage tanks filling when supply exceeds demand, and emptying when demand exceeds supply.

(ii)   Provide some back-up storage to allow continued operation of the system when supply is interrupted (e.g., pumps cannot operate because of a period without electricity supply).

Providing sufficient storage to fulfill the first function is essential, while providing some back-up storage is desirable.

The storage required to balance variations in demand depends on the supply regime (i.e., at what times do the pumps that supply the system operate) and the demand pattern. The supply regime depends on how pumps are powered. For instance, the supply regime for solar-powered pumps may be different from that of pumps powered from the electricity grid. The demand pattern will be roughly the same for most systems, reaching a peak in the morning, fluctuating during the day, reaching a secondary peak in the early evening, and dropping away to almost nothing overnight. If water is available throughout the day, the aim should be to provide storage equivalent to at least 12-hour demand on an average day.

Storage may be provided at the point of production, a central location within the school or to serve individual buildings housing sanitation blocks and kitchens. If the storage tank can be located within the envelope of the building structure, it will benefit from the heat generated by the building. Providing decentralized storage has another advantage in that the distribution system can be designed to carry peak day flows rather than peak hour flows.

Wherever possible, storage should be provided in elevated tanks, from which water can flow by gravity to taps. The storage tanks should be insulated so that stored water does not freeze in winter. Appendix 3 provides more detailed information on storage options and the detailed design of schemes based on preferred options.

Step 8: Selecting an Appropriate Sanitation Option

Basic sanitation choices concern the location of sanitation facilities and the type of toilet (water-flushed or dry). These choices are linked since the minimum requirements for WASH in schools require that dry toilets are located at least 20 meters away from dormitories and communal buildings. Washbasins should be provided close to toilets so that users can wash their hands after defecation.

For schools with a piped water supply, either existing or planned, water-flushed sanitation will be the preferred choice. This will allow sanitation facilities to be located within school buildings or in appropriately heated blocks linked to school buildings by enclosed walkways.

Water-flushed systems are often assumed to provide a higher level of service than dry toilets, because water-flushed systems reduce smells and problems associated with flies and other insects. Nevertheless, well-designed and maintained urine-diversion dry toilets and other forms of dry "ecological" toilets can, in theory, provide a similar level of service to water-flushed toilets. Achieving such a level of service, in practice, however, depends on user commitment to carry out tasks such as regularly covering the pit or vault contents with ash.

In most cases, pour-flush systems incorporating a water seal will provide a higher level of service than dry toilets, if the school has a functioning water supply system. Pour-flush toilets are normally preferable to cistern-flushed toilets because pour-flush toilets are simple without any moving parts, and require less water than cistern-flushed toilets, especially when lack of an adequate water supply makes it necessary to minimize the volume of flush water used required. The water requirement of pour-flush toilets will be around 3 liters per user per day for regular students, and 6 liters per user per day for students living in dormitories (Table 1). For a school with 400 students and teachers, of whom 200 are students living in dormitories, pour-flush toilets would require 1,800 liters of water per day, which is not a large volume.

The toilet is only the first link in the sanitation chain. Consideration of the subsequent links in the chain is essential if sanitation facilities are to be effective. Figure 4 shows options for containment, removal, and treatment of excreta and wastewater, distinguishing between water-flushed and dry toilet types. Solid connectors indicate activities that occur continuously or at short intervals, typically less than a week. Connectors shown with dashed lines indicate activities that are required at infrequent intervals.

Figure 4: Overview of Sanitation Options

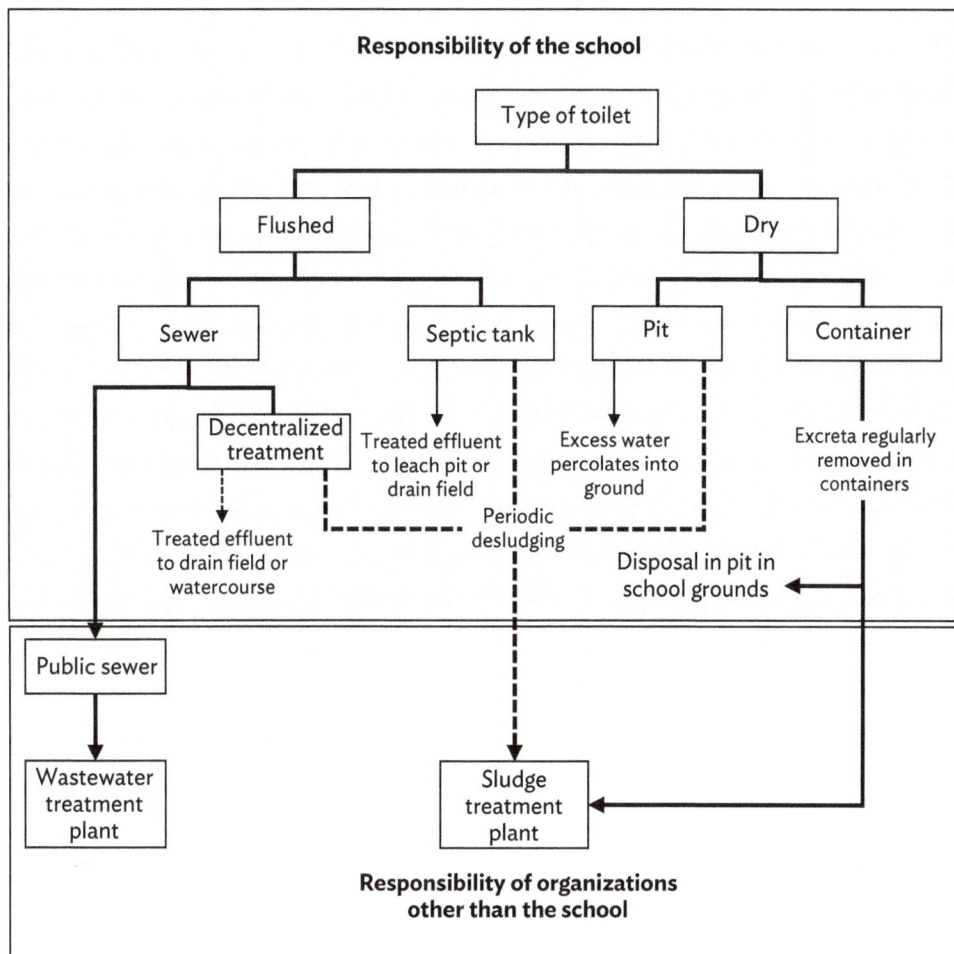

Source: Authors.

Where there is no public sewer, the simplest option for flushed toilets is to connect each toilet block to its own septic tank. Challenges with the septic tank option include the following:

(i)   Ensuring that the tank contents do not freeze, which normally requires that the tank location is below the depth to which the ground freezes. Locating the tank beneath a heated building will reduce the risk of freezing.

(ii)   The normal option for disposing of the tank effluent is to discharge treated effluent to a drain field or soakway. In extreme cold climates, the drain field or soakway should be located at a depth where it does not freeze during the winter.

(iii)   Periodic removal and safe disposal of accumulated and partly digested sludge, which will normally require a vacuum tanker to remove the sludge and transport it to a fecal sludge treatment plant.

If a *soum* center or *bagh* is too small to have its own vacuum tanker, the school should request the service of a vacuum tanker based at a larger town, such as the *aimag* center. To reduce costs, the aim should be for the tanker to empty all tanks in the small settlement at intervals of around 3 years. This requires cooperation with other institutions. The best option for fecal sludge treatment and disposal will be to provide simple drying beds at a suitable location.

Dry sanitation systems also retain fecal sludge on-site and require occasional removal and disposal of partly digested sludge. These types of sludge have higher solids content than the contents of septic tanks, and therefore cannot be removed using the suction hose of a vacuum tanker. Removal of sludge from a pit below a permanent sanitation block will thus be difficult. Although the possibility of replacing pits with containers located under toilets exists, sustaining such systems at the small scale required for school sanitation will be difficult, as suggested by the experience in Ulaanbaatar (Patinet and Delmaire 2015).[7] For these reasons, pour-flush toilets will be the preferred option, except where there is no short-term prospect of providing a reliable piped water supply to the school.

Step 9: Designing Hygiene Education Activities

The full benefits of WASH facility improvements will only be realized if users are educated on the importance of good hygiene and the proper use of the newly installed facilities. When planning hygiene education activities, it is therefore important to assess what people know and what they do at the planning stage. Hygiene education activities should start from an understanding of the existing situation. Any discrepancy between what people say they know and what they do should be investigated. Discrepancies may arise from a lack of facilities. For instance, while students may know that they should wash their hands with soap after defecation, they may fail to do so because there are no washbasins near the school toilets. This suggests that hygiene education should be linked with actions to provide WASH facilities, and that the WASH plan should consider hygiene education alongside proposals to improve WASH facilities.

Questions to ask when designing hygiene education activities include:

(i)   What should hygiene education activities aim to achieve?

(ii)   Who should they target?

(iii)   What methods should be used to deliver hygiene education messages?

(iv)   Who should be responsible for delivering them?

---

[7]   Trials using such systems to serve schools in Nairobi, Kenya have produced promising results (Bohnert et al. 2016).

A detailed guide for designing and carrying out hygiene education activities is found in section V. The most important aim of hygiene education is to ensure that students, teachers, and staff wash their hands with soap after defecation and before eating. Messages about the correct use of WASH facilities should be relevant to the facilities at the school. General messages or messages that refer to different types of facilities from those used by students will be of limited value.

Teachers are an important target group because they transmit knowledge to students and monitor progress in putting the knowledge into practice. The WASH plan should cover how the delivery of hygiene education messages will be monitored, and how the results will be evaluated.

Various standard approaches to hygiene education are available. Appendix 5 provides information on these approaches. At the planning stage, it is essential to consider the resource requirements of various approaches and choose an approach that is compatible with the resources available at the school. It is often advisable to simplify the approach to hygiene education by focusing on a small number of key messages, which should always include handwashing with soap; and using simplified versions of standard approaches, tailored to the school's needs and priorities and to the human and financial resources that are available at the school.

Step 10: Assessing Institutional and Financial Needs

Sustained improvements in school WASH will be possible only if systems are in place to ensure that, once provided, WASH facilities continue to deliver good quality services. The WASH plan should therefore include references to institutions, identifying the institutional arrangements required for effective facilities management. If the existing institutional arrangements cannot guarantee the long-term viability of proposed WASH systems, the plan should include realistic proposals to strengthen existing institutions and, where appropriate, introduce new institutional arrangements. When assessing institutional needs, the following specific points should be considered:

(i)   management needs of the WASH technologies proposed for the school,

(ii)   any requirement for cooperation and liaison between the school and service providers,

(iii)   institutional needs for hygiene education,

(iv)   operational costs of WASH facilities and services, and

(v)   support systems required to ensure that school level systems can function effectively.

When assessing institutional support needs, it is important to recognize that some schools are located in isolated *soums* and *baghs* and thus lack easy access to resources and services that are only available in *aimag* centers. Another point to consider is that the focus of schools is on education, and that their management systems may not be suitable for dealing with operational aspects of WASH services.

One response to the points above would be to develop strong *aimag*-level support systems for isolated schools, possibly involving regular visits to schools by personnel with appropriate technical and hygiene education knowledge and skills.[8]

---

[8]   In the 1980s, a mobile maintenance team was established to provide equipment repair services to technical schools in Sudan. The initiative started because maintenance teams based in individual schools lacked the equipment and expertise to carry out anything other than simple repairs and maintenance tasks. The Khartoum-based team travelled widely across the country to carry out maintenance and repair tasks.

Step 11: Preparing a Water, Sanitation, and Hygiene Plan

Once options have been assessed, preferred options should be consolidated into a WASH plan for the school. A WASH plan should cover the following:

(i)   the school's WASH needs (step 2);

(ii)   resources and constraints with particular reference to water sources and constraints created by lack of water, networked electricity, and institutional resources (step 3);

(iii)   proposed arrangements for extracting water from the preferred source, storing it, and delivering it to water outlets (sinks, washbasins, showers, and water-flushed toilets) at the school (steps 6 and 7);

(iv)   proposed arrangements for dealing with excreta and wastewater (step 8);

(v)   proposed hygiene education activities (step 9); and

(vi)   institutional and financial requirements of the plan, including any institutional strengthening proposal (step 10).

The plan should include drawings, sketches, photographs, and calculations for physical works. Supporting technical information can be included in appendixes. The plan should also include outline costings for the proposed WASH arrangements, prepared by qualified cost estimators. The assumptions underlying the costings should be clearly stated. If there is more than one feasible option, the plan should outline specific costs for each option. It is crucial to include operational costs and assess the availability of resources to cover these costs. The assumptions underlying the costings should be clearly stated. Appendix 6 provides further information on preparing cost estimates.

The WASH plan should be shared with various stakeholders, including intended users and concerned government agencies at the *soum*, *aimag* and, if appropriate, national levels. A meeting or workshop could be organized to present and discuss the plan with stakeholders, which might include the following:

(i)   presentation of the outline WASH plan;

(ii)   discussion on the outline WASH plan, during which any strong demand for changes or suggestions for improvements should be recorded;

(iii)   discussion on the roles to be played by different stakeholders in the implementation and subsequent management of the proposed WASH facilities; and

(iv)   exploration of options for financing the proposed WASH facilities and hygiene education activities.

Based on comments and suggestions received during the meeting or workshop, the WASH plan can be finalized.

IV. Improving and Managing Water, Sanitation, and Hygiene in Schools

With the water, sanitation, and hygiene (WASH) plan finalized, the stage is set for detailed design, construction, and management of WASH facilities.

A. Responsibilities for Designing, Constructing, and Managing Water, Sanitation, and Hygiene Facilities in Schools

Qualified engineers should prepare detailed designs, drawings, and costings for the proposed WASH schemes. Designs should follow engineering conventions and procedures, and should be in accordance with relevant Mongolian laws and standards. They should also reflect the minimum requirements for WASH in schools, while adequately considering climate and other relevant factors. Detailed information on technical aspects of providing water supply and sanitation in cold climates is in Appendixes 3 and 4.

Bidding and contract documents should follow Mongolian laws and regulations.[9] It may be appropriate to issue separate contracts for specialized tasks such as borewell drilling. Alternatively, main contractors can be encouraged to subcontract this work to specialist drilling contractors. Construction work will normally be awarded to suitably qualified contractors through competitive bidding procedures. The contractor will be required to work in accordance with the drawings and contract documents produced by the design agency or consulting firm, which should also be responsible for site supervision.

For small and relatively simple tasks, the option of awarding work to school management committees, using a form of community contract, might be considered. This approach is reported to have worked satisfactorily for schemes in Bangladesh (UNICEF and IRC 1998) and may be applicable for the implementation of school WASH improvement schemes in Mongolia. Government rules and regulations may need modification to allow community contracts.

Aimag-level and/or national-level technical departments should provide oversight. Where continuous presence of professional supervision engineers is not possible, one option will be to assign day-to-day site supervision responsibilities to people from the school community. These "community supervisors" should be selected through local structures such as school management committees. They should receive training from professional engineers, if possible using standardized training materials, and should report to the professional engineer with

---

[9]   For projects supported by development partners such as the Asian Development Bank, it may be necessary to add additional clauses and adapt the standard documentation in other ways to meet the requirements of the international partner.

overall responsibility for supervising the works. Community supervisors should be responsible for checking the following:

(i)   the quality of materials used in construction, which would require that they check the quality of components delivered to the site and carry out simple tests on materials to be used in the works; and

(ii)   key dimensions and components, to ensure that they are constructed as specified in drawings and specifications.

Monitoring of ongoing activities and evaluation of completed activities provide opportunities to learn from experience. The lessons from monitoring should be assessed and used to improve implementation while those from both monitoring and evaluation should be used to inform the design of future school WASH improvement programs.

The main responsibility for monitoring and evaluation will normally lie with those who finance and manage WASH improvement initiatives, including *aimag* or Ulaanbaatar education departments, relevant central government ministries, and, where appropriate, development partners. It is important that local stakeholders be involved in monitoring and evaluation activities, either as part of the monitoring and evaluation team, or by sharing their views on the adequacy of the facilities provided. When engaging with local stakeholders, it will be particularly important to obtain their views on problems and deficiencies and what might be done to overcome them.

Monitoring during construction should cover progress, construction quality, and construction cost. Problems and issues that arise during construction, together with the action taken to deal with them, should be recorded. After commissioning, monitoring must focus on the way in which the facilities are used.

Responsibility for the day-to-day management of WASH facilities falls on individual school principals and school management committees. Tasks to be allocated include regular cleaning of sanitation blocks, operation of pumps, routine maintenance of mechanical equipment, and checking for problems and deficiencies such as water leaks. Children may be encouraged to take responsibility for routine cleaning of the facilities allocated to them, with supervision and guidance provided by a teacher. Other routine operational tasks might be assigned to a specific individual, for instance, the school caretaker. Oversight responsibilities should lie with the school principal, aided as appropriate by the school management committee.

The following questions provide an overall framework for assessing management needs for water and sanitation systems:

(i)   What are the day-to-day operational needs of the new technology and what occasional maintenance and repair tasks will be required?

(ii)   What will be the management requirements of those tasks?

(iii)   What institutional arrangements will be needed to ensure the availability of the funds required to undertake essential operational tasks?

(iv)   Who should be responsible for day-to-day management and for providing oversight?

Management tasks for water and sanitation systems are divided into two broad categories:

(i)   "strategic" planning to ensure that the system functions well and will continue to do so in the future; and

(ii)   day-to-day management of routine operation and maintenance (O&M) tasks.

At the school level, the main concern will be with the management of routine O&M tasks, which is the main focus of this Guide. Strategic decision-making will mainly relate to identifying the need for future investment and will usually be the concern of higher levels of government.

It will be important to ensure that systems are in place to provide for continued operation, maintenance, and repair of WASH facilities and delivery of hygiene education programs. This may require rewriting of job descriptions; employment of additional staff to undertake O&M tasks; and contracts with private sector individuals and firms to provide operation, maintenance, and repair services.

Standard checklists should be developed for O&M tasks. This Guide provides indicative lists, but these are not exhaustive, and it will be advisable for each school to develop its own checklist, based on its own local situation. Students and teachers should be encouraged to report broken fittings and other problems. These problems should be recorded, as should the actions taken to deal with them.

Operational safety is very important. Nonspecialists should not attempt tasks that present any degree of danger to health. Appropriate safety apparatus should be available and used by those charged with undertaking these tasks.

More difficult maintenance and repair tasks will require input from *aimag-* or national-level institutions. Public utility service organizations, for example, should take responsibility for the regular testing of water and wastewater quality.

Implementation of these systems may require some institutional changes, such as rewriting of job descriptions and reassignment of responsibilities for maintenance tasks. Other possibilities include the introduction of some form of public–private partnership with individuals or organizations to undertake tasks associated with operation, maintenance, and repair.

Strategic aspects of management, including the provision of adequate funding, support systems, and occasional tasks such as undertaking major repairs and organizing desludging of septic tanks, will normally also be *aimag-* or national-level responsibilities.

The Ministry of Education, Culture, Science and Sports (MECSS) has the overall responsibility for managing schools' water and sanitation services. MECSS funds, plans, and oversees construction and major rehabilitation of school buildings, including WASH facilities, in consultation with aimag education departments. The main responsibility of MECSS is thus to ensure that the design of new WASH facilities takes account of operational needs while bearing in mind realities on the ground. Licensed engineers prepare the designs under MECSS supervision and take overall responsibility for construction supervision.

B. Routine Operation and Maintenance Tasks

For planning purposes, O&M tasks can be divided into the following categories:

(i)   routine, mainly operational, tasks;

(ii)   petty maintenance and repair tasks, needed fairly frequently, fairly simple and requiring only basic knowledge, skills, equipment, and finance; and

(iii)   occasional maintenance and repair tasks, needed much less frequently and generally more difficult to carry out than petty maintenance and repair tasks.

Routine operational tasks and petty maintenance and repair tasks will normally be the responsibility of the individual school or *soum*. Responsibilities for these tasks should be allocated to specific individuals. One possible arrangement will be to allocate tasks to a person who is already responsible for managing other aspects of school infrastructure, for instance, the person responsible for operating the school heating system. School management should take overall management responsibility, with the school principal providing oversight, together with the school management committee. Systems must be in place to ensure that these tasks are carried out promptly and effectively as and when required.

Every school should have an O&M manual, setting out the routine operational tasks and regular and occasional maintenance and repair tasks required for water supply and sanitation facilities. The manual should be as simple as possible, using diagrams and photographs where appropriate.

1. General Routine Operational, Maintenance, and Repair Tasks

Failure to keep sanitation blocks clean and in good condition will affect the image of the whole school and create serious health and safety issues. Pathogens will spread if fixtures, floors, and doors are dirty. Fouled toilet pans will attract flies. Unfortunately, the problem of unsanitary toilet and washing facilities is cyclical—dirty facilities attract more dirt, while inviting abuse and a reduction in care. To avoid such problems, it is important to introduce systems for keeping facilities clean as soon as sanitation blocks are commissioned. Routine tasks to be undertaken include

(i)   clearing any foul odor in and around toilet facilities;
(ii)   ensuring that floors are clean;
(iii)   ensuring that door handles are clean; and
(iv)   repairing broken and failed items, including light bulbs, doors, door handles, leaking taps. etc.

Overall responsibility for keeping sanitation blocks clean must lie with school management, particularly the school principal. Repair tasks should be assigned to the school caretaker, but it would also be good if responsibility for day-to-day cleaning tasks is devolved to students who use the blocks, under the supervision of a designated teacher.

The school should implement a system for reporting broken fittings and fixtures. Users should be encouraged to report problems, which should then be recorded by a responsible person, preferably a teacher. The responsible person can then issue an order for the caretaker to fix the broken item. To facilitate quick repair, each school should have a lockable store for replacement items including tap washers, taps, toilet pans, washbasins, and pipework fittings. Items should be issued from this store as and when required, and the transaction should be recorded. If good records are kept, it will be easier to justify any additional expenditure to replenish the stock in the store. All this requires money and it will be important for the school budget to include an allowance for the purchase of the items required for routine maintenance and repair of fittings and fixtures.

2. Routine Operational Tasks for Water Supply System

Box 2 lists routine operational tasks for water supply system components. Operational needs will vary from school to school, depending on the components included in the water supply system. Therefore, the list is only indicative of the operational tasks required, to be modified and expanded as necessary to suit the school's specific circumstances.

One person should be responsible for all the routine tasks listed in Box 2. In the event that routine inspection reveals problems that cannot be solved at the school level, *aimag* or Ulaanbaatar institutions should be contacted. If systems are to work well, these institutions must have the capacity to undertake nonroutine maintenance and repair tasks, and a supply of the parts and materials required to respond to commonly encountered problems.

Box 2: Routine Operational Tasks for the Water Supply System

Inspect wellheads at regular intervals, possibly once every month. Note any cracking and leakage, and repair minor problems.

Check for overflowing tanks, leaking pipes, and any problem with freezing components. Regular checks are required at no more than weekly intervals but preferably more often.

Read meters and pressure gauges, where they are provided, and record readings. If possible, do this daily and at the very least, once a week. Note any changes from the norm and explore possible reasons for those changes.

Check the operation of chlorinators.

Check taps and replace washers on all that are leaking.

Encourage users to report any drop in pressure at taps. This is an indicator that there is a leak in the pipework serving the tap.

Check operation of pumps and carry out basic maintenance tasks in accordance with manufacturer's instructions and guidance materials on pump maintenance.

Clean point of delivery devices as required by performance and in accordance with manufacturer's recommendations.

For self-contained systems, check operation of any disinfection system provided, carry out routine maintenance in accordance with manufacturer's recommendations, and arrange for repairs and major maintenance tasks to be carried out as and when required.

For chlorinated systems, regularly check chlorine residual at taps. Ideally, this should be done daily, if the school holds the necessary equipment. If daily testing is not practical, assess available resources and ensure that testing is carried out as often as possible.

Keep a record of stocks of any chemical required for water treatment and disinfection, and order replacement stocks in good time so as to ensure that there is no break in availability.

Check operation of valves by opening and closing them at least twice every year.

Inspect and clean water storage tanks at least once every year.[a]

[a] Washington State Department of Health has further information on tasks to be carried out daily, weekly, monthly, and at longer intervals.
Source: Authors.

3. Routine Operational Tasks for Sanitation Facilities

Box 3 is an indicative list of routine operational tasks for sanitation facilities. It provides a starting point for assessing O&M requirements but should be modified, with tasks added and/or adjusted as necessary, to suit the situation in individual schools.

Box 3: Sanitation Operational Tasks

DAILY TASKS

Check condition of toilets, preferably at least twice a day.

Clean toilet blocks as required, but at intervals of not more than 1 week.

Check operation of mechanical equipment, including pumps and air compressors used in wastewater treatment, and carry out any routine maintenance recommended by the manufacturer.

Report all systemic problems. This should be one of the tasks of those who check the condition of toilets. Problems should be reported to the person with overall responsibility for services, who in turn should report to the designated teacher or school management as appropriate.

OCCASIONAL TASKS

Clear blockages in pipework, as and where they occur. Each school should keep a set of the basic tools required for clearing blockages, such as drainage rods with properly engineered screwed ends and portable jetting equipment.

Desludge tanks and ponds. Septic tanks should be desludged every 2–3 years, depending on the space allowed for digestion and storage. This will often be a major task and will not be possible unless a vacuum truck or equivalent is available at the *aimag* headquarters, to be sent to *soum* centers as and when required. Fecal sludge should be disposed of safely, either at a wastewater treatment plant or by land disposal, carried out in accordance with relevant Mongolian legislation.

Source: Authors.

4. Safety and Occasional Maintenance and Repair Tasks

Safe practices are important but often ignored. Key safety rules are set out as follows:

(i)   Do not enter septic tanks and deep manholes without taking appropriate safety precautions.
(ii)   Do not work on electrical systems without first ensuring that the power is turned off.
(iii) Do not work with chlorine and other potentially hazardous substances in confined spaces.
(iv)   Deal promptly with any flooding resulting from leakage in pumping stations. Take particular care to ensure that wires carrying electricity are kept clear of temporarily flooded areas.
(v)   Promptly replace damaged electrical fittings.
(vi)   Replace open boarded covers to sanitation pits with properly designed and supported plastic or concrete slabs.

Specialist repair and maintenance tasks should only be undertaken by those who have been assigned to those tasks and have received appropriate training.

Another aspect of safety is the need to ensure that potable water is safe to drink. This requires regular water quality testing, followed by appropriate action if the water has unacceptably high levels of pathogens.

C. Funding for Water, Sanitation, and Hygiene Improvement Initiatives

It is not possible to sustain WASH-related activities over time unless funds are available to finance them. The point is of critical importance and should be reflected in school budgets. A United Nations Children's Fund (UNICEF) report stated that 1.7% of Mongolia's national budget for schools was spent on drinking water provision and wastewater management in 2011. The figure for kindergartens was less, at only 1.1%. In rural areas, the budget was used mainly to pay for the collection and transportation of water and for wastewater management. In urban centers, it was used mainly to pay water supply and wastewater management bills (UNICEF 2014). The long-term sustainability of WASH services depends on funding to cover not only such regular expenditures but also maintenance expenditure and occasional expenditure on equipment repairs and desludging septic tanks. In addition, funds will be required for hygiene education, which is sometimes overlooked.

The best way to ensure adequate funding for WASH-related tasks will be to include explicit provision for those tasks in school budgets. Separate items should be included for operational expenditure and for expenditure on hygiene education. Major repairs will be expensive and the need for such repairs will be difficult to predict. In view of this, funding for major repairs should be included in *aimag* budgets rather than those of individual schools.

Responsibility for allocating school budgets lies with *aimag* and Ulaanbaatar education departments. It will be important that the budget includes adequate provision for WASH operation, maintenance, and repair. The best option for ensuring this will be for the national government to issue guidelines setting out the provision that *aimag* education departments should make for schools in general and WASH facilities in particular, for operation, maintenance, and repair.

Provision should also be made for supply chain arrangements. This will require that the *aimag* education department keep an inventory of the WASH-related tools and equipment kept by each school, together with a supply of spare parts and replacement items. Systems to procure repair tasks that are beyond the capacity of individual schools and *soums* should be available at the *aimag* level. Systems will need to be developed to assess the need for repair. This will require some capacity development of *aimag* education departments, which currently have no staff with expertise in WASH facility construction and management.

It is possible that improved liaison with *aimag* technical departments might provide access to staff with relevant experience. Nevertheless, it is likely that the expertise that exists relates more to construction than it does to the ongoing operation, maintenance, and repair of facilities. Although Ulaanbaatar education department has staff with good experience in school and kindergarten building, including building of WASH facilities, there is still a need for increased emphasis on operational matters, including maintenance and repair.

V. Promoting Hygiene Education

In developing countries, fecal–oral diseases, including various forms of dysentery, cholera, and typhoid, are a major cause of mortality among children under 5 years of age. They also contribute to morbidity among older children and adults, reducing their ability to study and work and leading to poor educational outcomes and work performance. Clean water that is available to users in sufficient quantity to allow people to follow effective hygiene practices and good sanitation are essential components of any strategy to reduce the spread of fecal–oral diseases.

Some minerals and chemicals can be harmful to health if present in drinking water in concentrations that exceed maximum safe levels, as defined by the World Health Organization and/or national governments. For contaminants such as arsenic and fluoride, the effects on health may not be immediately obvious. High mineral levels can also affect the palatability of water. While this may not have direct health implications, it may have indirect impacts on health, in that people prefer to drink palatable but unsafe water over unpalatable but safe water.

A. Water, Sanitation, and Hygiene, and Fecal–Oral Diseases

Good water supply and sanitation facilities and services help prevent transmission of fecal–oral diseases—those diseases that pass from one person to another when a second person ingests a pathogen (disease-causing organism) excreted by an infected person.

Pathogens may be viruses, bacteria, protozoa, and various forms of intestinal worm. Almost all pathogens are contained in feces rather than urine.[10] Even a small amount of feces from an infected person contains enough pathogens to infect a second person who handles or touches it. Pathogens can pass from one person to another in the following ways:

(i)   Feces can contaminate water sources. Any person who drinks untreated or inadequately treated water drawn from those sources will ingest any pathogen that they contain.

(ii)   Pathogens may contaminate fields and the crops grown in them when
  (a)   people defecate in the open; and
  (b)   untreated feces and fecally contaminated wastewater are used to fertilize fields. Workers in the fields and those who eat crops grown in the fields are then at risk of coming into contact with the pathogens and ingesting them.

---

[10]   Schistosomiasis is an exception to this general rule, but it does not occur in Mongolia.

(iii)   Flies may settle on feces and then on food, carrying pathogens on their legs and bodies and contaminating the food, which is then eaten by people.

(iv)   An infected person may neglect to wash their hands after defecating, in which case fecal material containing pathogens may remain on their fingers and be passed on to another person, either directly by hand-to-hand contact or indirectly through food prepared or handled by the first person and eaten by a second person.

The F diagram in Figure 5 shows these transmission routes diagrammatically. Sanitary water supply and sanitation systems break these routes and prevent disease transmission, as shown diagrammatically in Figure 6.

Figure 5: F Diagram

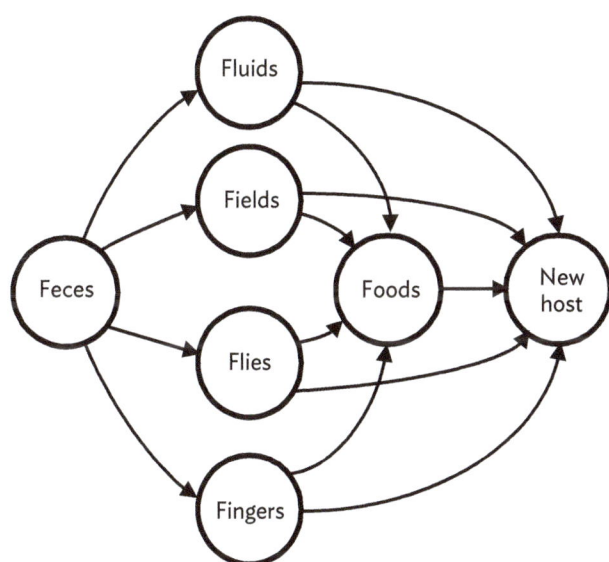

Source: Wagner and Lanoix (1958).

Figure 6: Water, Sanitation, and Hygiene Barriers to Disease Transmission

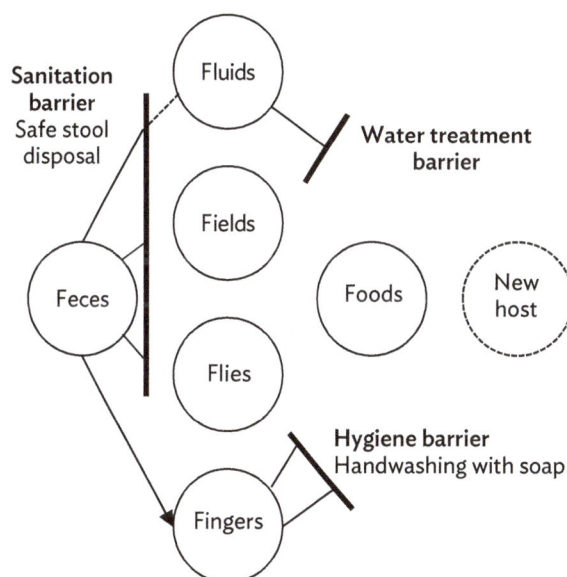

Source: Authors.

Treating potable water to kill pathogens breaks the direct transmission of pathogens that occurs when people drink polluted water. Almost all surface water sources, most springs, and many shallow groundwater sources are likely to be polluted and so will require treatment to remove pollutants and kill pathogens. Appendix 3 provides more information on water treatment options for removing and inactivating pathogens. Deep groundwater will usually be pathogen-free, although water from deep groundwater sources may still require treatment to deal with contamination that occurs during extraction, transmission, and storage.

Sanitation systems are designed to create a barrier between an infected person and other people through containment of excreta, ensuring that these do not reach water sources or food crops without appropriate treatment. This will only be achieved if the sanitation system encompasses the complete sanitation chain—from the toilet through storage, if required, and removal, transport, and treatment prior to safe disposal or reuse. The toilet should be designed to prevent flies from settling on fecal material and flying out to the household environment, where they may settle on food and contaminate it. Care must be taken to ensure that sanitation provides a complete barrier. An example of sanitation providing an incomplete barrier would be the situation in which the water table is high so that a pit latrine or septic tank drainfield penetrates it, leading to contamination of shallow groundwater. Even when provision is made for wastewater or septage treatment, the barrier may be breached.

The primary purpose of most conventional treatment systems is to reduce the organic content of the wastewater. However, few systems reduce pathogen concentrations to levels that are safe for drinking. Water discharged from a treatment system designed primarily to remove organic material may still contain pathogens, which will contaminate the water body to which it is discharged. In such situations, as in situations in which fecal material reaches water bodies untreated, potable water drawn from the water body must be treated to create the water treatment barrier shown in Figure 6.

Neither the sanitation barrier nor the water treatment barrier prevents transmission of pathogens via an infected person's fingers to food or directly to the hands of an uninfected person. Good hygiene, in particular, handwashing with soap, blocks this route. To ensure that the hygiene barrier is effective, a person should wash his or her hands both after defecating and before eating. An important point to note here is that good hygiene requires access to an adequate supply of water. Research in the 1980s showed that for hygiene purposes, the quantity of water provided is more important that the quality of that water. Its location is also important. A person is much more likely to wash their hands after defecating if water is available close to the toilet that they have just used.

B. Impacts of Minerals and Chemicals on Health

Both groundwater and surface water may contain minerals and chemicals, some of which are harmless while others pose a potential health threat. In particular, arsenic and fluoride can have serious long-term effects on health. People who drink arsenic-rich water over a long period may suffer from problems such as color changes on the skin; hard patches on the palms and soles of the feet; and skin, bladder, kidney, or lung cancer. Studies in Taipei,China found that long-term exposure to inorganic arsenic in drinking water increases the risk of contracting blackfoot disease, which causes severe damage to the blood vessels in the lower limbs, resulting eventually in progressive gangrene. There is evidence that long-term exposure to low levels of arsenic, such as those found in some groundwater sources, can increase the risk of developing diabetes (Bräuner et al. 2014).

While low levels of fluoride in drinking water can help to prevent tooth decay, higher concentrations will cause discoloration and pitting of the teeth. High fluoride levels in drinking water will eventually result in changes in the bones, a condition known as skeletal fluorosis. This can cause joint pain, restriction of mobility, and possibly increase the risk of some bone fractures.

When assessing source options for water supply schemes, the presence of arsenic and fluoride in water from each source should be investigated. Since the available studies show local variations in concentrations, it will be important to assess the quality of the particular source that is under consideration for each water supply scheme.

In some locations, high salinity and hardness levels may make existing water supplies unpalatable. Although water palatability is not directly related to health, it does have possible health consequences in that people may avoid drinking from an unpalatable but safe source, preferring another palatable but unsafe source. If this is the case, one objective to improve school WASH will be to ensure that water supplied to the school is palatable. Where the only source is either saline or very hard, this may require some form of treatment.

C. Hygiene Promotion

Hygiene promotion is the planned, systematic implementation of activities designed to ensure that people adopt activities and behaviors that prevent transmission of water, sanitation, and hygiene-related diseases. It should be a component of every school WASH improvement initiative. Hygiene promotion activities and behaviors relate to demand for improved water supply and sanitation systems and their subsequent use and maintenance, including appropriate storage and use of water in the home, hand and face washing, safe disposal of feces, and hygienic preparation of food.

Hygiene promotion involves a combination of three activities:

(i)   hygiene education directed at individuals, families, and communities to encourage adoption of hygienic behaviors and practices that help to prevent water and sanitation-related disease;

(ii)   service improvements, including both improvements in physical facilities such as water points and toilets and the development of outreach and support services; and

(iii) advocacy to encourage decision makers to adopt appropriate policies.

The three types of activity are linked. Advocacy will often be required to ensure that policy makers are committed to funding physical improvements in WASH facilities and hygiene education programs. Without this commitment, improvements in WASH services and health education efforts will be hard to sustain. People will not enjoy the full potential benefits of improved WASH facilities if they continue to act in an unhygienic way, for instance, failing to wash their hands after defecating. Similarly, hygiene education will have no effect if a lack of facilities or limited water availability prevents people from practicing the improved activities and behaviors to which they have been introduced through hygiene education. Children learn best by doing and will soon forget lessons if the recommended practices are impossible to follow in practice.

In sum, three interventions are needed to break the chain of fecal–oral transmission (see Figure 6):

(i)   action to ensure that students and teachers have access to an adequate quantity of water that is free of pathogens and harmful chemicals at the point of use;

(ii)   safe disposal of feces achieved through effective sanitation; and

(iii) handwashing with soap at critical times (after defecation and before eating).

D. Hygiene Education in Schools in Mongolia

The 2015 minimum requirements for WASH in schools specify the following:

(i)   Hygiene education, covering handwashing, tooth brushing, waste management, and proper usage of WASH facilities, should be part of the school curriculum.

(ii)   Hygiene education materials should be available and appropriate to the age of students and their development needs. Schools should play a role in raising public awareness on health-related matters.

(iii) Each school should have a fully trained staff member, designated as the focal point for WASH-related issues.

(iv)   Schools should support the establishment of clubs to raise awareness of WASH issues among students and encourage better hygiene practice.

(v)   An internal monitoring system should be established to assess changes in the attitudes and behavior of students following improvements in facilities and student exposure to hygiene education initiatives.

The challenge will be to ensure that these requirements translate into systems and activities at the school level. This requires that funding is available to develop training materials, train teachers, support the activities of hygiene awareness clubs, and carry out other activities associated with the hygiene education program. Financial provision for these tasks should be included in *aimag* education department budgets.

Overall responsibility for hygiene education activities will normally rest with the school principal. Trained teachers, school social workers, and school doctors should deliver hygiene education lessons and organize activities such as health clubs. Lessons should be participatory and provide opportunities for students to follow good practices. Ideally, each school should have at least two teachers trained to carry out these activities. The Ministry of Education, Culture, Science and Sports (MECSS) should prioritize the preparation of hygiene education training materials and should run regular courses, preferably at the *aimag* level. After attending an initial training course, teachers should attend hygiene education refresher courses at intervals not exceeding 5 years.

As with water and sanitation facilities, the approach to hygiene education should be standardized as far as possible. This requires action on the part of higher levels of government, which will normally assume overall responsibility for planning, designing, and funding hygiene education activities. Specific tasks to be undertaken at this level include

(i)   preparation of hygiene education materials;

(ii)   training of teachers to deliver lessons based on those materials;

(iii)   monitoring the content and quality of hygiene education activities in schools and, running refresher courses for teachers and others with responsibility for hygiene education lessons; and

(iv)   ensuring that basic materials, particularly soap, are available in schools.

In the short term, it may be necessary for MECSS to develop hygiene education materials, design programs, and train MECSS staff. The medium- to long-term aim should be to develop the capacity of MECSS to plan hygiene education programs and provide training to teachers and others with responsibility for delivering those programs.

The school principal will have overall responsibility for hygiene education activities in his or her school. Responsibility for hygiene promotion lessons and activities will normally be devolved to teachers. Every school should have at least one and preferably two teachers with training in hygiene promotion. For secondary schools, there should preferably be at least one trained teacher for every 200 students.

All students should receive training in good hygiene practices. The aim should be to ensure that students perceive good hygiene practice as the norm and follow it instinctively. To achieve this, lessons should include both demonstrations of good hygiene practice and opportunities for students to participate in following that practice. Key messages and practices should be repeated until students internalize them. This will require that hygiene education lessons and demonstrations are linked to implementation of physical WASH improvements so that there are no barriers to students following the good practices that they have learned.

One way to ensure that students are thoroughly versed in good hygiene practices and are encouraged to follow those practices is to form school health clubs. These are typically run after school with input from teachers and community health workers. For this approach to be successful, it will be important to ensure that teachers are given adequate training, in both the health lessons to be emphasized and on how to motivate students and plan and lead club activities.[11]

E. Requirements for Successful Hygiene Promotion

For hygiene promotion initiatives to be successful, they must identify the risk practices of the target audience and propose alternative practices that reduce the risk; they should be feasible in the local situation and are acceptable to the target audience. For schools, these risk practices will certainly include failure to wash hands with soap after defecation and before eating. In schools without piped water supply, risks associated with water transport, storage, and use should also be identified. Assessment of risk practices should start by identifying where pathogens might be present in the school environment and how those pathogens might infect people. Although the assessment does not have to be based on detailed scientific analysis, it does require a basic understanding of various types of pathogens, how they are transmitted, and how transmission might be prevented.

Because hygiene promotion is about human behavior, it needs to be culturally and socially sensitive. This means that hygiene promotion messages will only be accepted and adopted if they are

(i)   reasonable, in terms of the target audience's own rationale;

(ii)   recognizable to the target audience, which is a point that might seem obvious but is often overlooked; and

(iii)   acceptable to the target audience, in the sense that it fits with their view of themselves and their sense of self-esteem.[12]

If people in the target audience do not recognize or misinterpret a hygiene promotion message, they will not follow it, follow it incompletely, or follow unhygienic practices that are based on an incomplete or mistaken view of the message. Similarly, if they do not think that the message is reasonable and acceptable, they will ignore it.

These principles point to the fact that hygiene promotion in schools has to involve dialogue and interaction with teachers and students rather than one-way delivery of standardized messages. Since students already have ideas and beliefs, it is important that hygiene education efforts recognize the existence of these often deep-seated ideas and beliefs. Failure to do so may mean that changes in behavior resulting from hygiene promotion initiatives are short-lived, with students quickly reverting to old behavioral patterns.

Hygiene promotion efforts are more likely to be effective if they build on what students already know, recognize what they already think and do, and respond to what they want. Where existing beliefs create barriers to the adoption of good hygiene practice, hygiene educators should develop positive arguments for change, focusing on motivations that are important to students. Analysis based on formative studies in 11 countries found that the key motivations for handwashing were disgust, nurture, comfort, and affiliation. Fear of disease did not motivate handwashing except transiently during epidemic diseases such as cholera (Curtis, Danquah, and Aunger 2009).

---

[11]   Further information on health clubs, their potential advantages and possible drawbacks can be found in the case study developed by the International Water and Sanitation Centre (IRC) under the School Sanitation and Hygiene Education Global Sharing project financed by UNICEF (IRC 2006).

[12]   Based on Dudley (1993).

When developing hygiene education programs and materials, it is important to take account of the ways in which students' behavior may be influenced by the constraints that they face. Students do not wash their hands with soap if water is in short supply or available only from vendors at a high price, or soap is unavailable or expensive. One study found that people prioritize the use of soap for laundry rather than handwashing when soap availability is limited. The same study also suggested that washing hands with soap could involve a complex and time-consuming process of bringing together soap, water, and the vessels required to collect and pour water (Biran et al. 2012). Earlier studies had found an association between the presence of a washstand or water dispenser and both increased handwashing rates and reduced fingertip contamination (Biran, Tabyshalieva, and Salmorbekova 2005; Pinfold 1990). These findings point to the need to understand the barriers to the adoption of good hygiene practice and to take appropriate action to overcome those barriers. Information on methodologies for assessing existing beliefs and practices is found in Appendix 5.

F. Key Hygiene Promotion Messages

Hygiene promotion messages aimed at students and teachers should be simple, positive, and attractive to their intended audience. That suggests a need to focus on a small number of key messages, which emphasize the benefits to be gained from following good hygiene practice. It may be useful to use the "F" diagram (Figure 5) as a guide when framing messages. Whereas the most important hygiene promotion message is the need to wash hands with soap after defecating and before eating, messages could cover the following subjects:

(i)   the need to use water efficiently and minimize wastage;

(ii)   the importance of food hygiene, including the need to wash uncooked vegetables and fruit with safe water;

(iii) the need to use safe water for drinking and food preparation;

    (a)   In areas in which drinking water quality is poor, the basic message should be supplemented by messages on point-of-use treatment.

    (b)   Where water has to be transported in containers from one or more water sources, messages should include reference to the requirement to minimize the risk of water contamination during extraction from the source, transport, storage, and user access.

(iv)   the importance of collecting water in clean vessels without allowing the water to come into contact with hands, and carrying and storing water in covered vessels;

(v)   the dangers associated with drinking arsenic-contaminated water, with an emphasis on the need to ensure appropriate treatment (only where one or more potential water sources might have high arsenic concentrations); and

(vi)   messages relating to menstrual hygiene management (aimed at adolescent girls).

Hygiene promotion should also include messages about the correct use of water and sanitation facilities. These include the following:

(i)   the need to keep toilets clean, removing any fecal contamination from toilet bowls and pans (messages should emphasize everybody's responsibility for cleaning the toilet after use);

(ii)   the likelihood that blockages will occur if users attempt to flush bulky objects through water-seal toilets;

(iii) the practices to be followed when using direct-drop and urine-diverting dry toilet (UDDT) toilets; and

(iv)   the need to turn off taps after use and report leaking taps (this is particularly important during the winter when drops of water into drain pipes might lead to freezing problems).

Messages regarding the correct use of facilities should be specific to those found at the school. General messages or those that refer to different types of facilities from those used by the students will be of limited value.

The main focus of hygiene promotion efforts will usually be on facility users, primarily students. However, it is also worthwhile to address some promotion messages to those with responsibility for managing school WASH facilities. Such messages should cover the need to

(i)   maintain water systems and take action to discourage any practice and rectify any situation that might lead to contamination of water,

(ii)   ensure that simple water purification procedures are carried out as necessary,

(iii)   ensure that water storage tanks and containers are covered,

(iv)   ensure that sanitation blocks are reliably heated at all times, and

(v)   arrange for periodic removal and safe disposal of the sludge that accumulates in pits and tanks.

Messages aimed at those with WASH management responsibilities should emphasize the need to meet government WASH standards and minimum requirements. As with other hygiene promotion efforts, it is good to provide opportunities for dialogue designed to enable all concerned to develop a better understanding of the constraints that stand in the way of good hygiene practice.

G. System Needs for Hygiene Promotion

Hygiene promotion will only be sustained over time if effective systems exist to develop, manage, and deliver it. An effective hygiene promotion system requires the following elements over the long term:

(i)   trained hygiene promoters who are aware of the need to recognize people's existing beliefs, understand people's current practices, and investigate the constraints that prevent people from changing their practices;

(ii)   opportunities for those trained hygiene promoters to engage with students so as to facilitate learning on good hygiene practice;

(iii)   funding to support hygiene promotion activities;

(iv)   provision for monitoring ongoing hygiene promotion activities, evaluating their results, and using the lessons learned to improve the effectiveness of ongoing and future hygiene promotion programs; and

(v)   information and knowledge management systems that provide access to relevant information, sharing of experiences, training and resource materials, development of new approaches, and dissemination of findings.

Ensuring that these elements are in place will require action at the national level to develop guidelines, provide training opportunities, and assign adequate funding for hygiene promotion. These actions in turn, will need to be supported by adequate resources. This suggests that the starting point for action will be to convince policy makers of the need to assign resources to hygiene promotion in schools.

VI. Conclusion and Recommendations

Although there are existing materials on water, sanitation, and hygiene (WASH) in schools that provide useful and generally applicable guidance on school WASH facilities, they hardly address the specific physical and institutional challenges presented by the harsh climate and the small size and isolation of its rural settlements in Mongolia. As a result, technologies selected for WASH facilities in schools are often incompatible with existing infrastructure and services, climate conditions, and specific needs of schools. This Guide seeks to fill the gap by providing guidance on the processes to be followed and the decisions required to improve WASH in Mongolian schools.

For a long time, WASH in schools, especially in rural areas, has been given low priority in Mongolia. Beginning around 2010, poor conditions of WASH facilities and services in schools started to draw the attention of policy makers, which eventually led to the joint issuance of the minimum requirements for WASH in schools by the Ministry of Education, Culture, Science and Sports, Ministry of Finance, and Ministry of Health in 2015.

Meeting the minimum requirements for WASH presents a considerable challenge to many rural schools in Mongolia. One of the major constraints on school WASH improvements is the absence of an easily accessible water source or water availability, which makes it difficult to devise affordable WASH schemes. Even if a viable water source exists near schools, providing sustainable WASH improvements at an affordable cost remains a challenge in rural areas. Mongolia's harsh winter, with temperatures falling below freezing for long periods, requires WASH facilities to be provided with specific arrangements to ensure that water supply and sewer pipes laid close to the surface are heated and do not freeze. In the absence of centralized heating systems, which is mostly the case in many rural schools, heating costs are high.

Rural schools often rely on stand-alone, extremely basic WASH schemes because settlement-wide water supply, sewer, and heating systems are not available. Constant rural to urban migration reduces population growth in rural areas, with the result that rural *soum* centers have small populations, with many of their inhabitants continuing to live seminomadic lifestyles. The small populations of rural *soums* and the large distances between settlements increase the cost of WASH services, and cause significant logistical difficulties in sustaining those services. In small *soums* and *baghs*, critical services to properly manage WASH services in schools such as periodic desludging of septic tanks may not be available or expensive to organize.

Another major constraint is lack of funding to maintain WASH facilities. According to a United Nations Children's Fund (UNICEF) report (UNICEF 2014), only 1.7% of public expenditure on education was allocated to drinking water provision and wastewater management in 2011. In rural areas, the funding was used mainly to pay for the collection and transportation of water and wastewater management, while in *aimag* centers, it was used mainly to pay water supply and wastewater management bills. The long-term sustainability of WASH services critically depends on funding availability to cover not only such regular expenditures but also maintenance expenditure and occasional expenditure on equipment repairs and desludging septic tanks.

The lack of physical, institutional, and financial resources in many rural schools means that improving WASH services even to meet the minimum requirements is both difficult and costly. In this regard, this Guide aims to identify options for incremental improvements, designed to bring about immediate benefits, while paving the way for eventual full compliance with the minimum requirements and higher standards.

For instance, although the minimum requirements for WASH in schools stipulate that indoor water-flushed toilets should be provided for schools, and that outdoor latrines located at least 20 meters away from a dormitory or communal building should have insulation, ventilation, lighting, and handwashing points, outdoor toilet blocks without any enclosed and lighted access way and handwashing points continue to be built in rural areas. These toilet blocks need to be retained in the short term, while keeping in mind the longer-term objective of replacing them with hygienic toilets grouped in sanitation blocks and meeting the requirements. Some aspects of the requirements could be revised if improved forms of dry direct-drop toilets become available to allow their locations closer to school buildings than the currently specified 20 meters.

Responsibility for the day-to-day management of WASH facilities falls on the individual schools. It is important to ensure that systems are in place to provide for continued operation, maintenance, and repair of WASH facilities. This would require rewriting of job descriptions, employment of additional staff to undertake O&M tasks, and contracting private sector or community organizations to provide operation, maintenance, and repair services.

There is a need to develop capacity at the *aimag* level to gather and organize information, identify priorities, and commission WASH planning exercises. There should be effective coordination at the *aimag* level between the departments involved in planning and management of WASH services. There is also a need to develop systems to provide support to schools and organize tasks such as the periodic desludging of septic tanks, which are beyond the capacity of individual schools.

It may be necessary for *aimag* education departments to keep an inventory of the WASH-related tools and equipment owned by each school, together with a supply of spare parts and replacement items. Arrangements to procure repair tasks that are beyond the capacity of individual schools, *baghs,* or *soums* should be available at the *aimag* level. Systems also need to be developed to assess the need for repair. This would require some capacity development of *aimag* education departments that have currently no staff with expertise in WASH facility construction and management. While improved liaison with *aimag* technical departments may provide access to staff with relevant experience, existing expertise is likely to be related more to construction than to the operation, maintenance, and repair of WASH facilities.

Neither the sanitation barrier nor the water treatment barrier prevents transmission of pathogens via an infected person's fingers to food or directly to the hands of an uninfected person. Good hygiene, in particular, handwashing with soap, blocks this route. To ensure that the hygiene barrier is effective, people should wash their hands both after defecating and before eating. Good hygiene requires access to an adequate supply of water. Since people are much more likely to wash their hands after defecating if water is available close to the toilet that they have just used, the location of handwashing points is also important. Hygiene education, aiming to deliver a small number of key messages, should thus be part of WASH planning and improvement exercises. Support systems could be established at the *aimag* level to promote hygiene education activities at schools.

Finally, it is not possible to provide ongoing WASH-related activities—operation, maintenance, and repair of WASH facilities as well as hygiene education activities—unless funds are available to finance them. Adequate funding for WASH-related activities should be reflected in school budgets, which is the responsibility of *aimag* and Ulaanbaatar education departments. It is recommended that the national government issue guidelines requiring *aimag* and Ulaanbaatar education departments to allocate funds for the operation, maintenance, and repair of school WASH facilities.

APPENDIX 1
Physical Context of Mongolia

Section I introduced the challenges confronting many schools in rural Mongolia when improving water, sanitation, and hygiene (WASH) facilities. Among those challenges, physical conditions, particularly, climate, and the availability of water, often limit the options available for many rural schools to devise improved WASH schemes. Appendix 1 seeks to facilitate the understanding of the physical conditions essential to improve WASH services in schools.

Climate

The provision of water and sanitation services in Mongolia is influenced by its climate, in particular, its harsh winters and low rainfall.

Typical monthly temperatures range from 15°C to 25°C in July to –20°C to –30°C in January, depending on location. The lowest winter temperatures occur in the mountainous areas, particularly in the valleys between the Altai, Khangai, Khentii, and Khuvsgul ranges, where average January temperatures can range from –30°C to –35°C. The diurnal temperature variation is typically between 12°C and 18°C. Temperatures can fall to around –50°C in winter and reach almost 40°C in summer.

Some parts of the country experience permafrost conditions. Even where this is not the case, the ground freezes to a depth of several meters in winter. Figure A1.1 provides an overview of permafrost conditions in the country. The terms used in Figure 1 to describe permafrost have the following meanings:

(i) Continuous permafrost occurs where all ground remains frozen to a considerable depth throughout the year.
(ii) Permafrost is discontinuous where the ground thaws completely in some places during the summer.
(iii) Insular permafrost occurs where permanently frozen areas are surrounded by areas that thaw completely in the summer.
(iv) Sporadic permafrost exists where isolated areas of permafrost remain, although most of the ground thaws completely during the summer.

Figure A1.1: Extent of Permafrost Conditions in Mongolia

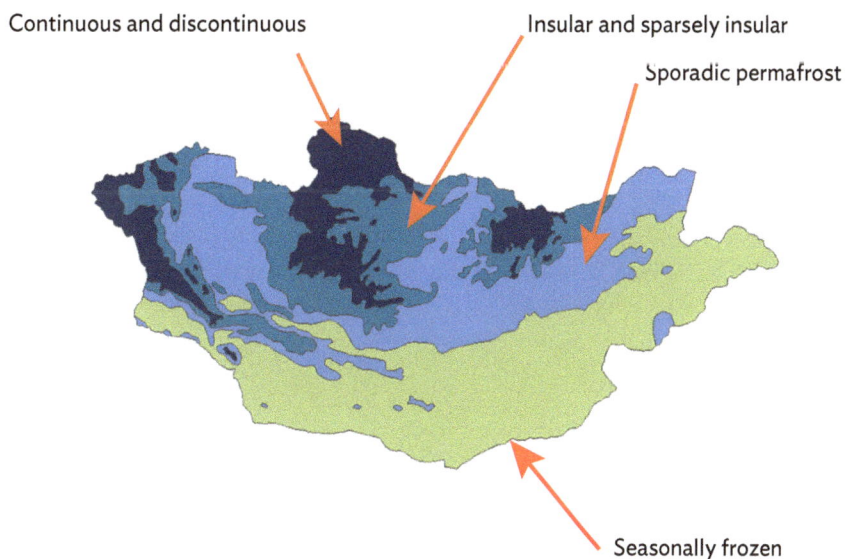

Source: National Snow and Ice Data Center. https://nsidc.org/sites/nsidc.org/files/images/data/ggd/geocryo_regions.png.

Permafrost conditions occur mainly in the north of Mongolia and in mountainous areas. The depth of permafrost can reach 100 meters (m) or more while the ground may thaw to a depth of 1.0 meter (m) to 4.5 m during the summer, depending on location. In areas with seasonally frozen ground, the ground may freeze to a depth of 1.6 m to 5.0 m, contingent on the severity of the winter and the location.

The extreme winter temperatures have important implications for the design of water supply and sanitation systems. In permafrost areas, deep groundwater is only available below the permafrost. Water may exist at shallow depths above the permafrost during summer. However, water available at shallow depths has a high probability of contamination by pathogens and chemicals leaching from pit latrines, leaking sewers, and other sources of contamination. Overall, accessing clean water in permafrost areas is normally both difficult and expensive. Moreover, in both permafrost areas and those with seasonally frozen ground, measures to prevent pipes and tanks from freezing in winter are required. Many water and sanitation facilities in locations across Mongolia have failed because the designs did not adequately consider the effects of extreme cold on their performance.

Rainfall amounts in Mongolia are generally low. Ulaanbaatar's average annual rainfall is about 380 millimeters (mm). Monthly precipitation peaks in July at almost 80 mm and falls below 5 mm from October to April. Rainfall patterns are generally similar in other parts of the country, although total rainfall in the north and east exceeds that in Ulaanbaatar, while total rainfall in the dry southwest is much lower.

The following general points on water supply should be noted, all of which have an impact on the availability of water to supply schools:

(i)   long dry periods rule out rainwater harvesting as a water supply option,
(ii) low overall rainfall amounts mean that there are few surface water sources other than along main rivers, and
(iii) low rainfall amounts reduce recharge of aquifers.

Mongolia is a predominantly sunny country, on average experiencing around 250 sunny days each year. The possibility of solar energy should be investigated when considering options for powering pumps in settlements without an electricity grid connection. Average wind speeds in some areas are high enough to suggest that wind could also be a possible power source.

Water Availability

Water availability should be a central concern in planning for improved WASH in schools, because every water supply system requires a source that is both adequate and reliable. Many washrooms and toilets in schools are not in use simply because there is no adequate source of water to supply them. Box A1.1 summarizes the general situation surrounding water availability in Mongolia.

Box A1.1: Water Availability in Mongolia

The main sources of water in Mongolia are rivers, streams, lakes, springs and groundwater.

There are over 4,000 rivers, of which around 20 rivers are large, mostly flowing north flowing north toward the Arctic Ocean. Few rivers and streams exist in the drier south of the country. Rivers in the west and southwest of Mongolia flow south into the desert, where some feed lakes or systems of lakes. In such cases, the final lake in the system is normally saline. All rivers freeze completely during winter, except for a few large ones.

A 2004 study (Altanzagas 2006) identified around 9,600 springs in the country, of which 1,484 were dry at the time of the study. The study also estimated that around 10% of Mongolia's population were using spring water.

Groundwater accounted for 80% of all Mongolia's freshwater consumption in 2010 (ADB 2014). In areas with limited rainfall, groundwater may be fossil water, which receives little or no recharge from the surface.

Sources: B. Altanzagas (2006) and ADB (2014).

Water Quality

Two aspects of water quality impact upon the choice of source for school and settlement-wide water supply systems: microbiological quality, in particular, the likelihood that pathogens are present; and physical and chemical quality. The microbiological quality of deep groundwater is usually better than that of shallow groundwater, including springs and surface water. The physical and chemical quality of groundwater can affect both the safety and the palatability of water.

The water quality issues in Mongolia relate to the following:

(i) **Arsenic.** This occurs in groundwater in the southern Gobi region and may be an issue in adjacent regions.

(ii) **Fluoride.** High levels have been recorded in many wells in the southern Gobi region, especially in the northeastern part of Omnogovi Aimag and in Dundgovi and Dornogovi Aimag.

(iii) **High total dissolved solids concentrations. These** are usually associated with salinity. High total dissolved solids (TDS) concentrations affect the palatability of water, at worst making it undrinkable, particularly in southern and southwestern Mongolia, including parts of the southern Gobi region. Further information on acceptable TDS levels is given in section II.

(iv) **Hardness.** This is mainly caused by calcium and magnesium carbonate. Most of Mongolia's groundwater is hard to a greater or lesser extent. At extreme levels, this may influence palatability and make groundwater unsuitable as a drinking water source. At lower levels, hardness causes scaling on utensils and pipes.

High levels of arsenic and fluoride can both adversely affect health, as explained in section V.

Assessing Water, Sanitation, and Hygiene Needs

Section III outlined the process and priorities in planning for improved water, sanitation, and hygiene (WASH) in schools, based on assessments of school WASH needs, resource availability, and constraints. Appendix 2 provides guidance on information collection and analysis for the assessments, especially for school management, *aimags,* and Ulaanbaatar education departments. This appendix covers information collection needs and procedures relating to

(i)   the school and the number of students, teachers, and staff to be served;

(ii)   existing and planned WASH facilities;

(iii)   standards and regulations;

(iv)   climatic conditions;

(v)   institutional responsibilities and capabilities;

(vi)   water availability;

(vii) power availability; and

(viii) hygiene knowledge and practice.

Information Type 1: School, Students, Teachers, and Staff

To estimate the demand for WASH services, information is required on (i) the total number of students and students living in dormitories, teachers, and school staff; (ii) the location and uses of existing and planned school buildings; and (iii) the layout of existing buildings.

1.   Student, Teacher, and Staff Numbers

Information on student, teacher, and staff numbers is needed to determine how much water should be supplied to the school; and how many toilets, washbasins, and showers will be required. A distinction must be made between regular students and students living in dormitories since the needs of the two groups will differ: students living in dormitories use more water and need access to facilities that are not used by regular students. School management, under the leadership of the school principal, should gather copies of written records of the number of students, divided into regular students and students living in dormitories, as well as the number of teachers and school staff, broken down by gender.

To assess future needs, information is also required on the likely increase in the number of staff and students over time. There is evidence that schools with improved WASH services attract students. This possibility must be taken into account when assessing staff and student numbers.[1]

The procedure for estimating future student and teacher numbers should involve the following steps:

(i)   Decide the planning design horizon (typically 20–25 years in the future).

(ii)   Gather information on the present number of students, teachers, and staff.

(iii)   Obtain information on the general population growth rate at the *aimag, soum* (or *bagh*), or Ulaanbaatar district levels, disaggregated by gender.[2] Assess the growth rate for the number of regular students and students living in dormitories, teachers, and staff at the school by taking account of the general population growth rate, plans for new classrooms, etc.

(iv)   Calculate student, teacher, and staff numbers at the planning design horizon, using information on current numbers and the assumed growth rate.

2. Location and Uses of Existing Buildings

Information on the location and uses of existing buildings is needed to determine where WASH facilities should be located and how many facilities of each type are required for each location. For initial planning, encourage teachers and students to prepare a sketch showing the location of school buildings, any existing water supply and sanitation facilities, and potential sites for future facilities. This sketch can then be used as a basis for discussion and later refined with the help of specialists. Google Earth can be used in the mapping process, if the quality of images is good. Good quality images can be reproduced to an exact scale (for example, 1:500) and used as a base for identifying school buildings and facilities. Figure A2.1 is taken from a Google Earth image showing part of Bayan-Uul Soum in Govi-Altai Aimag.

Figure A2.1: Public Buildings, School, and Kindergarten in Bayan-Uul Soum

Bayan-Uul Soum in Govi-Altai Aimag. *Part of Bayan-Uul Soum, showing public buildings including school and kindergarten* (photo from Google Earth).

---

[1]   An evaluation report on the school WASH project supported by UNICEF and AusAID in Khuvsgul Aimag referred to *soum* leaders and school principals who reported that herder families are willing to send their children to the schools where WASH facilities had been improved under the project.

[2]   Population data, including annual population growth rates, are available from the website of the National Statistics Office of Mongolia (http://www.1212.mn/).

3. Layout of Existing Buildings

Other subjects for investigation are layout and heating arrangements of existing buildings. Questions to ask include the following:

(i)   Is there spare that could be converted for use for washing and toilet facilities?

(ii)   If not, where could stand-alone sanitation blocks be located?

(iii) Does the heating system use hot water pipes and radiators, and if yes, could the system be extended to serve new sanitation blocks?

(iv)   Is there room in the roof space to accommodate water storage tanks?

Information Type 2: Existing and Planned Water, Sanitation, and Hygiene Facilities

For planning purposes, the following information is required:

(i)   number, location, and condition of existing facilities (toilets, washbasins, showers, etc.);

(ii)   water availability—is it available from taps, or does it have to be carried from a remote source, etc.?;

(iii) any operational problems of existing facilities, for instance, nonfunctional sinks and flush toilets—what are the reasons for facilities that have fallen into disuse?; and

(iv)   user views on existing facilities and services.

Information on existing facilities should cover those that are both currently in use and abandoned. In the case of abandoned facilities, it is important to understand why the facilities have failed. Such understanding illuminates constraints and difficulties that need to be taken into account when developing proposals for new facilities.

When assessing existing facilities and services, the following questions should be asked:

(i)   Does the school have a piped water supply? If so, is it operational?

(ii)   If water has to be carried from outside, what arrangements exist for transporting and storing it?

(iii) What washing facilities exist, where are they located, how much water is provided, and is hot water available?

(iv)   What toilet facilities exist, what form do they take, and what is their condition? What provision, if any, exists for toilet users to wash their hands after defecating?

(v)   Are there any facilities that are no longer used? If so, what are the reasons that they were either never used or fell into disuse?

(vi)   How is the school heated in winter? Does each building have its own boiler or are all buildings heated from a central boiler? If a central boiler exists, will it be possible to lay water and possibly wastewater pipes alongside the heating pipes?

Along with assessing existing facilities and services, the present per capita water use should be assessed. If there is an existing piped systems supplied by pumped groundwater, it may be possible to measure the borewell output by recording the time that an overhead tank or container of known capacity takes to fill. Multiplying the recorded output by the time that the borewell pump operates during the day will yield the volume of water produced each day. Further dividing this by the number of users will then give the average per capita water production. This calculation method does not distinguish between regular students and students living in dormitories.

There are two options for obtaining more information on the water use of each of these groups. The first is to observe and record water facilities use over a typical day. The second is to encourage students and teachers to self-record their water use over a typical day. It is better to use both methods so that the results from each can be cross-checked.[3] The results of the water use assessment should be compared to the standards given in section II. It would then be possible to calculate the increase in water quantity and the increased number of facilities required to bring provision up to the required standards.

Information Type 3: Standards and Regulations

Information on the current standards and regulations of the Government of Mongolia is provided in section II.

Information Type 4: Climate and Its Effect on Ground Conditions

The key point to consider regarding climate is the temperature and its effect on ground freezing. Two questions should be investigated:

(i)   Do permafrost conditions apply and if so, what is the depth of the permafrost and to what depth does the ground thaw in summer?

(ii)   If permafrost conditions do not apply, to what depth does the ground freeze in winter?

These questions can be answered using secondary information (Appendix 1) and local information obtained from *aimag*- and Ulaanbaatar-level institutions. Where permafrost is present, boreholes must extend to below the lowest level of permafrost. Elsewhere, pipes running outside buildings must either be located below frozen ground or heated in some way to prevent freezing.

Information Type 5: Institutional Responsibilities and Capabilities

Institutional responsibilities are defined by the law. However, the systems and structures that exist in reality may vary from those specified in the law. These variations often stem from a discrepancy between the capacity required for formal arrangements and the much lower capacity that is actually available.

The strengths and weaknesses of institutional systems can be gauged through the following questions:

(i)   Are responsibilities for various aspects of WASH management, including hygiene education, clear?

(ii)   Who has responsibility for what?

(iii)   How are responsibilities allocated in practice—do actual management arrangements differ from the formal arrangements?

(iv)   What capacity exists to manage operation and maintenance (O&M) of water and sanitation services?

(v)   In the event that management capabilities are limited, what options exist for strengthening them?

---

[3]   Monteiro and Gomes (1998) has a description of a self-recording exercise.

It is worthwhile to seek answers to these questions in relation to specific WASH-related tasks. Table A2.1 shows a possible proforma for identifying responsibilities, problems, and possible solutions relating to individual tasks.

Table A2.1: Proforma for Recording Water, Sanitation, and Hygiene Responsibilities

Task	Official Responsibility	Actual Responsibility	Problems	Options for Overcoming Problems
Manage water supply system				
Operate water supply system				
Maintain and repair water supply system				
Manage washing and sanitation facilities				
Clean washing and toilet facilities				
Maintain and repair washing and toilet facilities				
Empty toilet pits when full				
Plan for future service provision				
Organize hygiene education activities				
Deliver hygiene education				
Provide soap for washrooms				

Source: Authors.

The list of the tasks above is not necessarily complete and some tasks, for instance, emptying toilet pits, may not be required for all schools. Nevertheless, filling out a proforma like Table A2.1 helps define operational responsibilities and identify areas in which improvements will be required to ensure effective management of a WASH improvement scheme.

Information Type 6: Water Availability and Quality

The following questions should be asked when investigating possible water sources:

(i)   From where does the school obtain water at present? What information is available on the adequacy of this source in terms of both quantity and quality?

(ii)   Is piped water available near the school? Does the piped water system have enough capacity to support an improved supply to the school?

(iii) What other potential sources exist? How far are they from the school? What information is available on the likely yield and quality of water from each potential source?

If a new or improved source is required, it is likely that information on yield and quality is limited, particularly in the case of groundwater. In that event, the following points will serve as a guide in the search for suitable groundwater supplies:

(i)   Groundwater is more likely to be found under valleys than under hills. The presence of water at the surface in springs and seeps indicates the presence of groundwater but not necessarily in large quantities.

(ii)   Existing wells can provide useful information on the aquifer and the potential for supply from new wells. Investigation of existing wells should establish their location, the depth to the static water table, the rate at which water is extracted, and information on the types of rock penetrated by the wells, which might possibly be available from the drilling records. Chemical analysis of the water extracted from existing wells would provide information on the quality of the water found in the aquifer.

(iii) Geological records help to put findings on topography, the location of springs and seeps, and the yield and quality of borewell water into perspective. In areas with uniformly bedded sedimentary rocks, aquifers may be extensive. If water is available in fissured rock, projecting findings from existing wells to other locations would be more difficult and problematic.

Professional advice should be sought when investigating groundwater availability. However, local knowledge from people around the school should also be tapped.

Information on possible sources should be included in the design report in tabular format, showing the type of source (shallow or deep groundwater, spring or surface water); key quality indicators (presence of *E. coli* and concentration of important chemical impurities); an estimate of its yield, and its distance from the school. Table A2.2 is an example of a table compiling information on possible sources.

Table A2.2: Sample Table Recording Existing Water Sources

Serial no.	Type of Source	Distance from School (km)	Approximate Yield (l/sec)	Comments and Observations
1	Depth of shallow well (m)			Describe quality of water and method used to transport water to school
2	Depth of deep borehole (m); water table depth (m)			Any issues with chemical quality
3	Spring			Describe any measure that have been taken to protect the spring from pollution
4	Lake			Describe method used to transport water to school

km = kilometer, l/sec = liter per second, m = meter.
Source: Authors.

Existing boreholes could also provide information on the likely depth and yield of the groundwater reservoir from which they draw. Even if a borehole is no longer operational, it may still be possible to plumb it to determine the depth of the water table.[4] Records and discussions with local people could provide further information, although the information on yield obtained from conversations may not be accurate.

Topography and geology also provide clues as to where groundwater, in particular, relatively shallow groundwater resulting from local run-off, might be found. Groundwater resulting from local run-off is more likely to be found under valleys than under hills. Outcrops provide information on the direction of dip of underlying strata, which can be used to assess where water-bearing strata lie underground. Springs indicate the presence of groundwater. The challenge, however, is to assess the direction and spread of the aquifer that is feeding the spring. Vegetation may provide another clue to the location of shallow groundwater. Where groundwater lies underneath, vegetation should be more abundant. Geological maps and records would also provide information on potentially water-bearing strata.

---

[4]   Cunningham and Schalk (2011) has further information on measuring the static water level in a borehole.

If the methods outlined here do not provide good information, it will be necessary to commission detailed specialist technical surveys involving techniques such as measurements of electrical resistivity. If additional geophysical information is required, it will also be necessary to engage specialists to carry out the investigations required to obtain the information.

Information Type 7: Power Availability and Options

Most water supply technologies and some sanitation technologies require energy to power pumps and/or treatment processes. When considering technology options, it is therefore important to consider their power requirements, as well as the options for providing power supply, and the cost of the supply. Where there is a reliable electricity supply network, connection to the network will be the best option.

Questions to be asked about power availability include the following:

(i)   Is the school connected to a centralized electricity network?

(ii)   Is the supply continuous or near continuous?

(iii) In the event that there is currently no connection to a centralized electricity supply, is such a connection foreseen in the near future?

School management should be able to answer the first and second questions, whereas electricity supply authorities can provide information on any plans for future extensions to the electricity supply system.

Most schools in settlements without networked electricity supply depend on diesel-powered generators. This is an expensive option, particularly if the settlement is remote, making the cost of transporting fuel high. Other sources of power include solar, wind, and water-powered hydroelectric plants. Photovoltaic solar power potential exceeds 4 kilowatts per meter squared per day (kWh/m^2/day) for most of Mongolia and exceeds 5 kWh/m^2/day in southern and southwestern regions of the country.[5] Because of the low rainfall, the potential exists throughout the year, although output will be greater in summer than in winter. Wind power also has good potential throughout most of the country (National Renewable Energy Laboratory 2001).

Solar power is arguably more flexible and has lower installation costs than wind power, but both are worth investigating in situations where there is no centralized electricity supply. Organizations specializing in renewable energy should have information on solar radiation and average wind speeds throughout the year. Any information on the performance of existing solar and wind installations should be collected, analyzed, and used to assess whether actual facility performance approaches the performance that is theoretically possible. The results of the investigation could provide inputs into national databases on solar and wind power resources.

Information Type 8: Current Hygiene Knowledge and Practice

Information on existing hygiene knowledge and practice is necessary to prepare a plan for improving hygiene practices based on the actual rather than assumed conditions.

---

[5]   The photovoltaic power availability map of Mongolia is from SOLARGIS (World Bank 2019).

Investigation of current hygiene knowledge and practice could center on the following questions:

(i)   What do students know about hygiene?

(ii)   What do teachers know about hygiene?

(iii) What do students actually do to maintain basic hygiene?

(iv)   What do teachers actually do to maintain basic hygiene?

(v)   If there are differences between what students and teachers know and what they do, what are the reasons for those differences?

The answers to the last question, particularly reasons for differences between knowledge and practice, may point to constraints on improving hygiene practice. Improvements in practice will require action to overcome those constraints, which might relate to physical deficiencies and/or deeply ingrained patterns of thought. For example, even if students know what constitutes good practice, they cannot improve their hygiene practice because facilities are not in place to apply that knowledge. Simple factors such as the lack of soap in washrooms may prevent students from carrying out the basic discipline of handwashing with soap. More information on methods for gaining a better understanding of existing beliefs and practices is found in Appendix 5.

Knowledge and practice of teachers can be explored through guided group discussions (focus group discussions). Understanding how teachers think is important because of their role in transmitting knowledge to students and monitoring their progress in putting the knowledge into practice.

In addition to assessing the current situation in individual schools, efforts should be made to identify any hygiene education materials and programs available at the national level. It is likely that such materials were produced in the course of specific projects and programs, often with support from development partners. Information should also be gathered on the extent to which such materials are used. If they are not in general use, it is useful to explore the reasons for the limited use.

Assessing Additional Information Needs

The previous sections set out procedures for gathering information by collecting secondary data, including those from observation, satellite images, and consultation with key informants. Although these procedures usually enable schools and *aimag* or Ulaanbaatar education departments to gain a reasonable understanding of existing conditions, proposed plans and important constraints, information on some factors with an important bearing on WASH-related decisions may still be far from complete.

In some cases, it may be possible to infer missing items of information from knowledge and experience of similar situations. However, inference will not be enough to provide hard evidence on factors such as the presence and potential yield of aquifers, and the ability of various forms of decentralized electricity generation to provide the power required to drive pumps and other equipment. Where initial investigations reveal important information gaps, additional studies will be necessary and adequate allowance for such studies should be made, if WASH-related decisions are to be truly evidence-based. Specialized guidance should be sought on the scope of the studies required and the budget needed for those studies.

Assessing Water Supply Options

Appendix 3 provides detailed guidance on the planning and design of water supply schemes to serve schools. It is mainly intended for professional engineers and architects with responsibility for planning and designing school water, sanitation, and hygiene (WASH) schemes. This appendix covers the following:

(i) objective of improving WASH in schools in compliance with the standards and regulations set by the Government of Mongolia, regarding the type and location of water points to be provided, per capita water demand, water quality, and continuity of supply;

(ii) calculation of water requirements of the school, based on per capita water demand with appropriate allowance made for monthly and daily variations in demand and water losses;

(iii) assessment of possible water sources;

(iv) assessment of water supply options, in particular, whether a school should have its own water supply, or should be provided from a water supply designed to serve the whole settlement (e.g., soum or aimag center, or subcenter of ger areas) where the school is located;

(v) assessment of options for extracting water from the source and delivering it to the school, and how should water extraction pumps be powered;

(vi) assessment of water treatment options, including whether treatment will be required, and how and where should treatment be provided;

(vii) assessment of water transmission, distribution, and storage options; and

(viii) assessment of user interface options.

Planning for school water supply requires attention to the whole supply chain—from the source, to the taps, and to other water outlets that deliver water to the school. It should start with determination of the volume and quality of water required and the ways in which it is accessed by students, teachers, and school staff. This requires information on the number of regular students and students living in dormitories, teachers, and school staff at the design horizon; and per capita water requirements, as specified in the minimum requirements for WASH in schools.

Once the school's water needs have been established, attention must turn to the options for meeting those needs. The first task is to identify water sources with the potential to meet the school's water needs. If a public water supply system with spare capacity exists, the best option will be to serve the school from it. If there is no public supply system, other sources must be explored. Except where there is heavy mineral contamination, groundwater will normally be the preferred water source. If the only source is at some distance from the settlement where the school is located, the school should work with other stakeholders to develop a new public supply system for the whole settlement. If a local source is available close to the school, a scheme to supply the school alone could be considered, while carefully examining its operational needs.

Water in a system that serves only the school should be disinfected, regardless of the source. Surface water, shallow groundwater, and deep groundwater contaminated with minerals will all need additional treatment to remove impurities. Treatment may be provided prior to distribution or close to, or at the point of delivery. Treatment prior to distribution is preferable for public systems. Otherwise, local treatment at or close to individual water outlets may be preferable.

Most water supply systems require storage to balance variations in supply and demand. Where possible, the storage should be located at a height that allows water to flow to taps and other water-using devices by gravity. Although conventional water distribution systems provide storage at central locations, the best option for schools will usually be to provide storage close to water outlets (taps, washbasins, connections to toilet cisterns, etc.). Storage should be inside buildings to benefit from the heating provided for the buildings.

Systems should be designed to operate during winter conditions. That means, water mains should be either placed below the greatest depth of frost penetration or heated and insulated in some way to prevent freezing.

Objective

The overall objective of improving school water supply services is to ensure access to water services for students, teachers, and school staff that protect health and environment and enhance the convenience and privacy of their users. Water supply services should meet the standards and regulations set by the Government of Mongolia (section II) in every respect. The standards concern the location and number of water outlets provided, per capita water requirements, and the quality of water (guidance on how to assess WASH needs in accordance with the standards is provided in section III and Appendix 2).

Calculating Water Requirements of Schools

The amount of water required from the system can be estimated following the steps below:

(i)   Calculate the average daily demand for water.
(ii) Apply a peak factor to the average daily demand to calculate the maximum demand to be expected on any one day (the peak daily demand).
(iii) Add an allowance for leakage and wastage.

Calculation of the average daily water requirements of a school requires information on the number of regular students, teachers, and school staff; the number of students staying in dormitories; the standard per capita demand for each group; and the allowance to be made for leakage and wastage. The number of regular students, teachers, staff, and students living in dormitories should be estimated at the design horizon for the water supply system, typically 20–25 years into the future. The simplest way to calculate future student numbers is to obtain information on current student and teacher numbers and apply a growth rate equivalent to the average population growth rate for the area where the school is located.

Average per capita water demand can be calculated using the figures given in Table 1 (section II). In most cases, it will be appropriate to use the figures given for systems incorporating pour-flush toilets. The average water demand for a school is given by equation (1)

$$D_s = N_r \times q_r + N_d \times q_d \qquad\qquad (1)$$

where

> D_s is the average daily demand in liters,
> N_r is the number of regular students and teachers,
> N_d is the number of students staying in dormitories,
> q_r is the regular student per capita water requirement in liters, and
> q_d is the dormitory student per capita water requirement in liters.

For a school with 500 students and teachers, comprising 350 regular students and teachers, and 150 students living in dormitories, using the figures given in Table 1 and assuming cistern-flushed toilets, equation (1) gives an average daily water demand of 350 x 25 + 150 x 37 = 14,300 liters. Using pour-flush toilets rather than cistern-flushed toilets will reduce the average daily water demand to 350 x 11 + 150 x 28 = 8,050 liters.

In most countries, water demand varies through the year, reaching a peak during hot dry periods. It also varies significantly during the day, reaching a peak during the morning when people are washing and preparing food, and falling to almost zero overnight. The standard way of allowing for varying demands is to apply peak factors to the average daily demand, which are

(i)   the "peak day" factor, which is used to size production facilities, transmission mains, and storage tanks and reservoirs;

(ii) the "peak hour" factor, which is used to size distribution mains when there is no storage within buildings served by the system. If storage is provided within sanitation blocks and kitchen buildings, the external water distribution system should be designed to carry peak day flows.

The values of these factors depend on local conditions. Typical values in temperate climates are 1.5 for the peak day factor and 2.5 for the peak hour factor. The United States Environmental Protection Agency suggests higher values for small systems in cold climates: 2.3 for the peak day factor and 4.5 for the peak hour factor. These figures apply to systems that are continuously pressurized. Where the system is pressurized for only a few hours each day, the peak factor may be higher.

For most schools in Mongolia, it is likely that peak day demand is suppressed because supply capacity is limited. At the design stage, judgment is required to identify appropriate peak factors. It is important that assumptions on peak factors are clearly stated in design calculations so that their accuracy can be assessed in the light of subsequent operational experience.

All water supply systems lose some water during transmission, storage, and distribution. This loss may be due to both leakage from water mains and storage tanks, and from wastage from open taps and overflowing storage tanks.

Leakage rates vary greatly, depending on the condition of water mains and storage tanks. The leakage rate for a recently built system, with welded polyethylene pipes and good quality plastic storage tanks, should be less than 5% of average daily demand. Leakage rates from old systems with pipes and tanks that have suffered some damage due to freezing conditions might exceed 50%. Good system management should reduce wastage to a minimum. In water-scarce situations, which is the norm in Mongolia, the aim should be to reduce leakage to a minimum and eliminate wastage. Measures to be taken to achieve this aim are discussed later in this appendix.

The water production capacity required for a school is given by equation (2):

$$PC = D_s \times (P_d + LF) \hspace{4cm} (2)$$

where

> PC is the production capacity in liters per day;
> D_s is the average daily demand for water;
> P_d is the assumed peak day factor;
> LF is an allowance for leakage, represented as a proportion of average daily demand.

For the school with 350 regular students and teachers, and 150 students living in dormitories, assuming cistern-flush toilets, a peak day factor of 1.5 and leakage equivalent to 15% of the average daily demand, the required water production capacity would be 14,300 x (1.5 + 0.15) = 23,595 liters per day, or 23.595 cubic meters per day.

Assessing Possible Water Sources

The biggest challenge for many school WASH schemes will be to identify a viable water source reasonably close to the school. Possible water sources include:

(i)   an existing water main;
(ii)   surface water (streams, rivers, and lakes);
(iii)   springs; and
(iv)   groundwater.

Rainwater collection is not a viable option in Mongolia because of the low annual rainfall and the long dry season experienced throughout the winter months (Appendix 1 has more information on Mongolia's climate). It is therefore not considered here.

Where a school is within the service area of an existing piped water supply system, the piped system will usually be the best supply option. The only exception is where demand exceeds the system capacity, resulting in low pressure and an intermittent water supply. Even if this is the case, options for upgrading the supply system should be explored to allow the school to be served.

If there is no piped water supply, other sources must be identified and assessed, including existing sources. For instance, where water for the school is carried from a shallow well fitted with a bucket or hand pump, the existing source will be inadequate in both quality and yield. Similarly, where the school is dependent on vehicle transport of water from a remote source, the arrangement will also clearly be inadequate because it is expensive and can only deliver a limited volume of potentially contaminated water.

However, both arrangements might provide useful information on the possibilities for identifying an improved source. If, for example, an existing well is located in a natural depression, and the presence of shallow groundwater is confirmed, it is possible that a greater yield is achieved by a well drilled to a greater depth. In the case of a vehicle transport from a remote source, the possibility of piping water from the source should be assessed, while recognizing that the pumping costs may make the scheme too expensive to be viable.

Water from surface sources such as rivers and lakes almost always contains pathogens and thus requires treatment to render it safe for drinking. Springs and shallow groundwater are also likely to contain pathogens and their yield may be limited. Deep groundwater is much less likely to contain pathogens and should therefore be preferred unless it contains unacceptably high concentrations of chemical contaminants. Box A3.1 provides further information on the source options to be investigated when there is no existing public water supply.

Box A3.1: Brief Review of Water Source Options

Surface water. This option is possible for a school located close to a stream, river, or lake but it has two disadvantages. The source is likely to freeze during the winter. Small streams and shallow lakes that freeze throughout their depth during winter will not provide water year-round. For larger lakes and rivers, the freezing will not extend beyond surface layers and water will still be available below the ice. The challenge in such cases will be to design an intake that continues to function through the winter, does not silt up, and is protected from damage by ice blocks carried in the flow during the spring thaw. Another disadvantage is that water from a surface source is likely to be contaminated and will require treatment before use for potable purposes.

Springs. These are fed by shallow groundwater, which is more susceptible to fecal contamination than deep groundwater. A 2004 study found that of the 9,600 springs in Mongolia, 1,484 were dry, and that about 10% of Mongolia's population, particularly poor households, use spring water (Altanzagas 2006). The availability of good quality spring water varied across all regions. For example, flow rates in Khovd and Dundgovi Aimag were generally good, while those in Orkhon and Bayankhongor Aimag were generally poor. Water quality, in terms of clarity and lack of turbidity, was generally better in rural *aimags* than around Ulaanbaatar. Detailed investigations of 127 springs (99 from rural areas and 28 from Ulaanbaatar, chosen because their yield was at least 3 liters per second, were usable throughout the year, and produced palatable drinking water) revealed that 77.2% of the springs studied had no protection. The remaining 22.8% had minimal protection in the form of wooden, iron, and stone fences. Even if the water feeding the spring is uncontaminated, contamination occurs at the spring itself unless the spring is improved and protected.

Groundwater. It may be shallow (i.e., groundwater that extends to within about 7–8 meters of the surface and is accessible to a suction pump located at ground level) or deep and, in the case of deep groundwater, confined or unconfined. Some existing shallow wells are located at some distance from the settlements that they serve. Their shallow depth and lack of protection mean that the water that they produce is likely to be contaminated.

If there is no impermeable barrier between the groundwater table and the surface, the groundwater is unconfined. An impermeable barrier, such as a layer of clay, confines the groundwater, preventing it from rising to its "natural" level. Confined groundwater may be under pressure, in which case it rises to a higher level in any borehole that penetrates it. In some parts of Mongolia, deep groundwater is "fossil" water—water that has been present in the ground for centuries or even millennia without recharge from the surface. The challenge with fossil water is to estimate the volume of water stored in the aquifer to establish whether it is sufficient to meet the long-term needs of the school or settlement served by the water supply scheme.

Groundwater is often present under a lake, river, or stream, and is available even in the winter when the water at the surface is frozen. This may be particularly important in discontinuous permafrost areas, where the only ground free of permafrost could lie beneath a river or deep lake.

Source: Authors.

The challenge is often to obtain reliable information on the potential yield from water sources, particularly from groundwater sources. The *aimag* or Ulaanbaatar public utility service organization may have relevant information and, in particular, any stream flow records and/or records of the diameter, depth to water table, and yield of existing boreholes, which might throw light on the likely yield of sources that are under consideration for the school.

The yield of groundwater sources depends on the aquifer characteristics, borehole diameter, length and position of the well screen, and capacity of the borehole pump. One report on the southern Gobi region quotes typical borehole yields of between 1 and 15 liters per second (l/sec). Yields from "traditional" shallow wells consisting of an excavation, around 2–3 meters (m) deep, lined with rocks or concrete blocks, would be much lower. The yield of springs is normally low, and thus, a spring should only be considered as a water source if its yield is at least 0.5 l/sec. Box A3.2 introduces methods for gauging spring flows.

Box A3.2: Methods for Gauging Spring Flows

The most accurate option for gauging small flows is to record the time taken to fill a container of a known capacity. This method is used when sufficient fall is available downstream of the spring to allow the flow to be directed through a pipe or channel projecting over a point where the container can be placed to receive the flow.

When the flow is too large to use this method, and/or when the available fall is limited, flow is measured using a V notch cut into a metal plate, fabricated from 10 to 16 gauge steel or galvanized sheet iron. The formula for the flow through a 90° V notch weir is $Q = 1400H^{2.5}$, where Q is the flow measured in liters per second and H is the height from the "V" of the weir to the upstream water level measured in meters. The water level drops through the weir, and therefore, the upstream level should be measured at some distance upstream of the weir.

Sources: World Health Organization. Flow Measurement and Control. *Fact Sheet* 2.9. http://www.who.int/water_sanitation_health/hygiene/emergencies/fs2_9.pdf; and Hydromatch. *Flow Estimation for Streams and Small Rivers*. http://www.hydromatch.com/sites/default/files/downloads/DIY-flow-measurement-guide.pdf.

Assessing Water Supply Options

Figure A3.1 sets out a logical procedure for carrying out an assessment of water supply options. The options that are technically and institutionally simple should be examined first. Only if local conditions do not allow adoption of a simpler option, should more complex options be considered.

Figure A3.1: Procedure for Assessing Water Supply Options

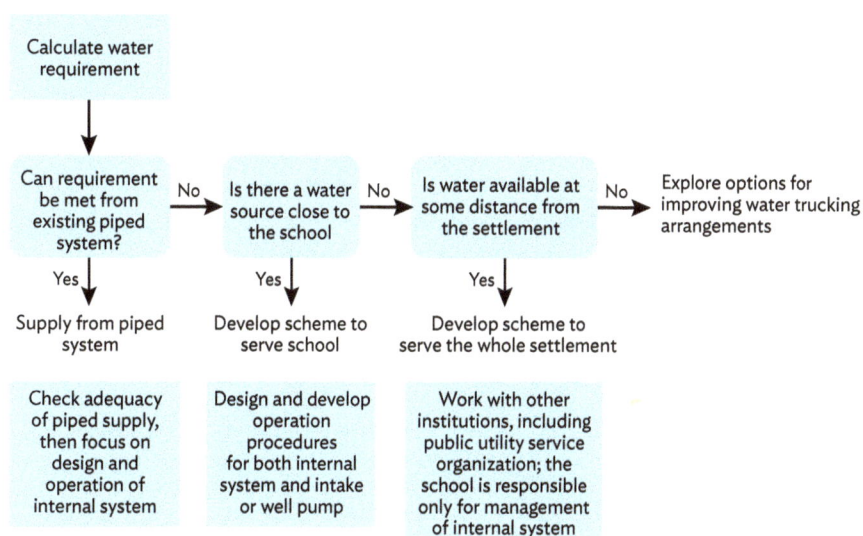

Source: Authors.

Figure A3.1 assumes that it is difficult and expensive to introduce a stand-alone scheme serving only the school when the water source is at some distance from the settlement where the school is located. The best option in such circumstances will be to develop a scheme to serve the whole settlement. This will require cooperation between several organizations and funding beyond the resources available to the school authorities. Securing the required funding and reaching agreement on roles and responsibilities for implementation and subsequent operation of the water supply scheme will often take time.

If funding is not available in the short term, the best strategy will be to identify and implement immediate improvements that can be made using the existing water source, while working with other concerned organizations to develop an improved water supply scheme to serve the whole settlement in the longer term. Immediate improvements might include action to protect existing shallow wells from pollution and introduce larger water carrying vehicles to increase carrying capacity and reduce transportation costs. These should be seen as temporary measures, to be superseded as soon as constraints on the implementation of more permanent improvements can be overcome.

Surface water will rarely be the best option for small settlements because of the difficulties encountered in keeping the intake frost-free during winter and the likelihood of pollution. One exception might be small *soums* located alongside large rivers and lakes that do not freeze completely and have no alternative source of water.

Assessing Water Extraction Options

The arrangements for water extraction depend on the nature and location of the water source. Possible water extraction options include

(i)   surface water intake,

(ii)   spring capture, and

(iii) borewell fitted with electrically or diesel-powered pump.

Surface water intakes and spring capture installations are normally followed by a pump or pumps to lift the water to storage tanks in school. The only exception is where the intake or spring capture is located at an elevation that allows gravity supply to school. Pumps in boreholes are normally designed to pump water directly to the school storage tanks.

1. Surface Water Intake

Rivers and lakes in Mongolia freeze for 5–6 months every year, with the depth of ice reaching between 0.8 m and 3.2 m (Punsalmaa, Nyamsuren, and Buyndalai 2004). If a river intake is to work throughout the year, it must take water from a depth below the greatest depth of freezing. The depth of freezing depends on the flow and the water conditions, as explained in Box A3.3.

Box A3.3: Water Freezing Mechanisms

The depth to which a water body freezes depends on location and flow. Lake ice freezes from the surface downward because slight differences in density mean that the coldest water is at the surface as temperatures drop to around freezing point.

Turbulent water freezes more slowly than still water and the freezing process is different. Where the flow is turbulent, as is the case in a fast-flowing river or stream, warmer and colder water constantly mix with one another so that there is no temperature gradient. The lack of a temperature gradient means that the first stage in freezing is for small needles of ice, known as frazil ice, to form through much of the depth of flow. As the temperature drops further, the frazil ice gradually consolidates into an ice layer. As ice forms at the surface, it inhibits the transfer of energy between the cold air and the warmer underlying water so that the rate of ice formation starts to drop. Nevertheless, ice thickness will continue to increase, albeit at a gradually reducing rate, as long as severe freezing conditions prevail.

Analytical methods to calculate ice depth, taking account of air temperature and the period over which freezing conditions have prevailed, are available. However, assessment of the likely maximum thickness of ice cover should be based on local knowledge and field measurements, especially when assessing a potential surface water source.

Source: Authors.

Smaller rivers and streams may freeze over their whole depth. In such circumstances, one way to access water is to construct a small dam downstream of the take-off point to contain the flow, and increase the depth sufficiently to leave some unfrozen water beneath an ice covering. The dam should incorporate a sluice or a pipe controlled by a penstock to allow draining and desilting of the impounded water.

The cost of intake construction is not proportional to intake capacity since a relatively small increase in intake size can result in a much larger increase in capacity. In view of this, and taking account of the need to limit disturbance of the aquatic environment, the intake capacity should carefully be determined to meet the likely ultimate water demand.

The delivery pipework should include a flow meter and a pressure gauge. The flow meter provides information on cumulative flow and preferably includes provision for reading the instantaneous flow. It should be installed in a frost-free location, either in a frost-protected building or an accessible underground meter chamber.

The intake of surface water must be located some distance above the bed to minimize the chance that silt is drawn into the water supply system. Figure A3.2 shows a simple arrangement, suitable for use in a scheme to meet the smaller demand generated by a single school or small *soum* center.

Figure A3.2: Arrangement for Lake or River Intake

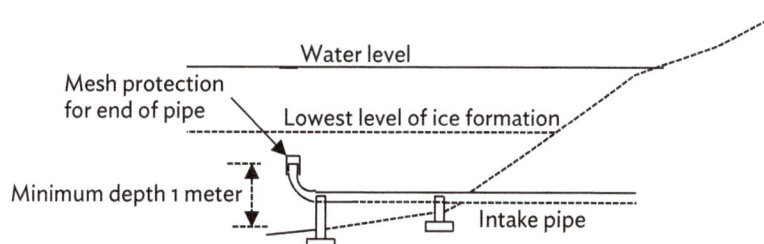

Source: Authors.

The end of the intake pipe curves upward and ends in a bellmouth. If installed in a river, the intake must be protected from scour, siltation, and fouling by objects carried in the river flow. To minimize scour and siltation problems, the intake should be located on a straight reach of the river channel, where neither erosion nor deposition is likely to occur. Other options for guarding the intake pipe against damage include

(i)   burying the horizontal section of pipe and providing a protective crib around the intake pipe where it emerges from the riverbed; and

(ii)   locating the intake in a recess built into the bank of the river.

River and stream flows decline in the winter months, when the only flow into streams and rivers is from groundwater located beneath frozen ground. Even when the surface flow ceases completely, groundwater flow may continue below the frozen river. Such groundwater may be tapped using shallow boreholes fitted with submersible pumps.

A better method to collect groundwater may be to construct an infiltration gallery. Infiltration galleries are horizontal conduits, laid in permeable water-bearing earth below the water table, and designed to collect groundwater and direct it to a collection well, from which it can be pumped. The infiltration gallery itself consists of a perforated pipe laid in a bed of gravel, laid about 1.5 m below the lowest groundwater level. Normally, several such pipes are laid to converge on the collection well, as shown in Figure A3.3. A case study of infiltration gallery schemes in Alaska is in Box A3.4.

Figure A3.3: Diagrammatic Representation of an Infiltration Gallery

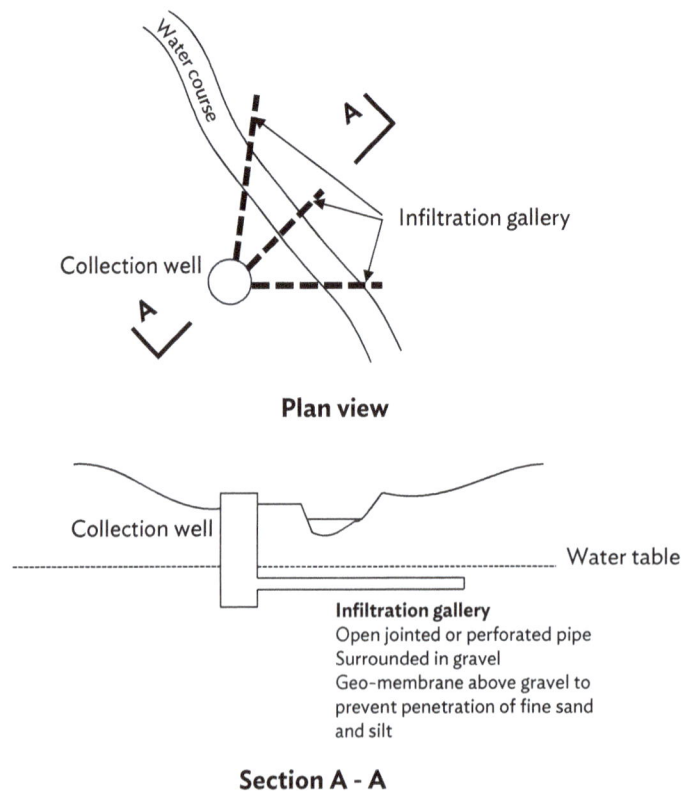

Plan view

Section A - A

Box A3.4: Case Study of Infiltration Galleries in Alaska

Some small settlements and industrial concerns in Alaska obtain water from infiltration galleries, located under streams and rivers. Examples of infiltration galleries are the Golovin settlement scheme in the western part of Alaska and the Green's Creek Mine scheme in southeastern Alaska. Infiltration galleries have been used as a source of water supply for communities in Alaska for over 50 years.

In areas with highly mineralized groundwater, infiltration gallery water may be the only water that is free of pathogens and with an acceptable mineral content. Unlike surface water sources, infiltration galleries are not affected by ice formation during the winter or damage caused by ice break-up during the spring.

The Green's Creek Mine infiltration gallery replaced an earlier stream intake that produced unacceptably turbid water and had been subject to damage and blockage caused by the bed load in the stream. Its infiltration gallery connects to the sump used for the earlier intake. Water is pumped to the mine from this sump (Figure A).

Figure A: Section through Infiltration Gallery Installation

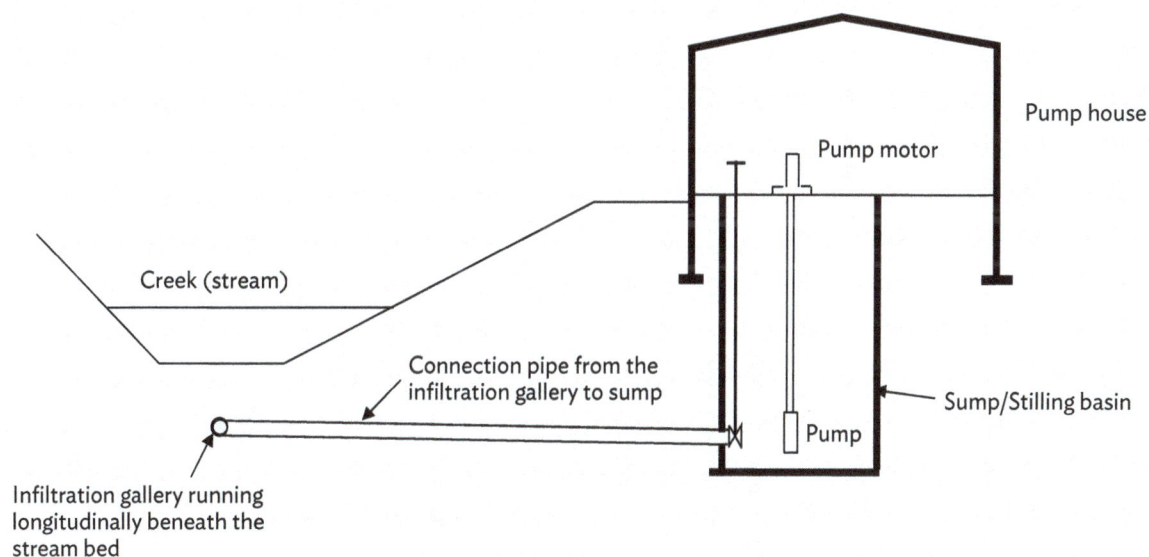

The infiltration gallery consists of a perforated screen forming part of a horizontal pipe laid beneath the bed of the stream in a trench filled with broken stone and gravel. The screen is constructed using two concentric slotted steel pipes, which form an annulus that is filled with small glass beads. This arrangement occupies less space than a conventional filter pack and the beads are less subject to encrusting and fouling than conventional filter packs (Figure B).

continued on next page

Box A3.4 *continued*

Figure B: Section through Screen

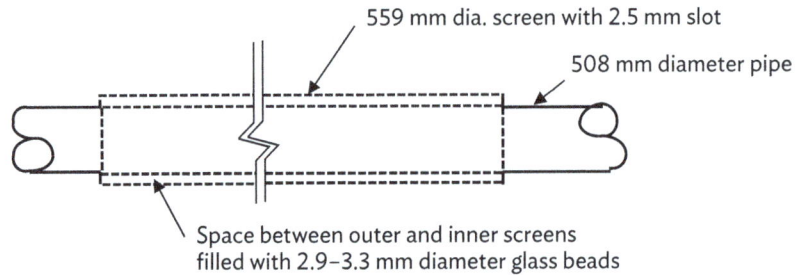

559 mm dia. screen with 2.5 mm slot

508 mm diameter pipe

Space between outer and inner screens
filled with 2.9–3.3 mm diameter glass beads

mm = millimeter.

A small diameter "airburst" pipe runs inside the screened pipe. An air compressor delivers air bubbles through the pipe as and when required to remove any debris that is fouling the screen.

An aqua barrier, a portable water-inflated cofferdam, fabricated from industrial grade vinyl-coated polyester, was used to dam the Green's Creek stream to allow construction of the infiltration gallery.

Source: Hanna, Sundberg, and Hayden (2016).

Source: Authors.

If a *soum* or school is close to a stream with a good fall, as is the case in some mountainous parts of Mongolia, the infiltration gallery can be located upstream of the *soum* or the school so that water can be supplied by gravity (Figure A3.4).

Figure A3.4 Possible Concept for Gravity Supply

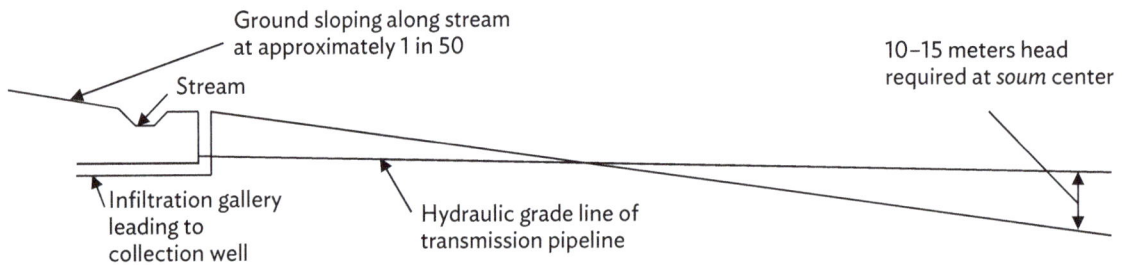

Ground sloping along stream
at approximately 1 in 50

Stream

10–15 meters head
required at *soum* center

Infiltration gallery
leading to
collection well

Hydraulic grade line of
transmission pipeline

Source: Authors.

2. Spring Capture

As indicated in Box A3.1, over 75% of springs accessed for water supply in Mongolia have no protection, with the remaining springs having minimal protection. Inadequately protected springs produce contaminated water, often in small quantities. This suggests a need to develop springs to reduce the risk of contamination and increase the yield. The standard approach to spring development is to dig back into the "eye" of the spring, replace some of the earth excavated from the eye with gravel, and provide a connection from the improved eye to a collection chamber, which is normally referred to as a spring box. Figure A3.5 shows one possible arrangement for collecting water from a spring and feeding it into a piped water supply system.

Figure A3.5: Arrangement for Typical Improved Spring

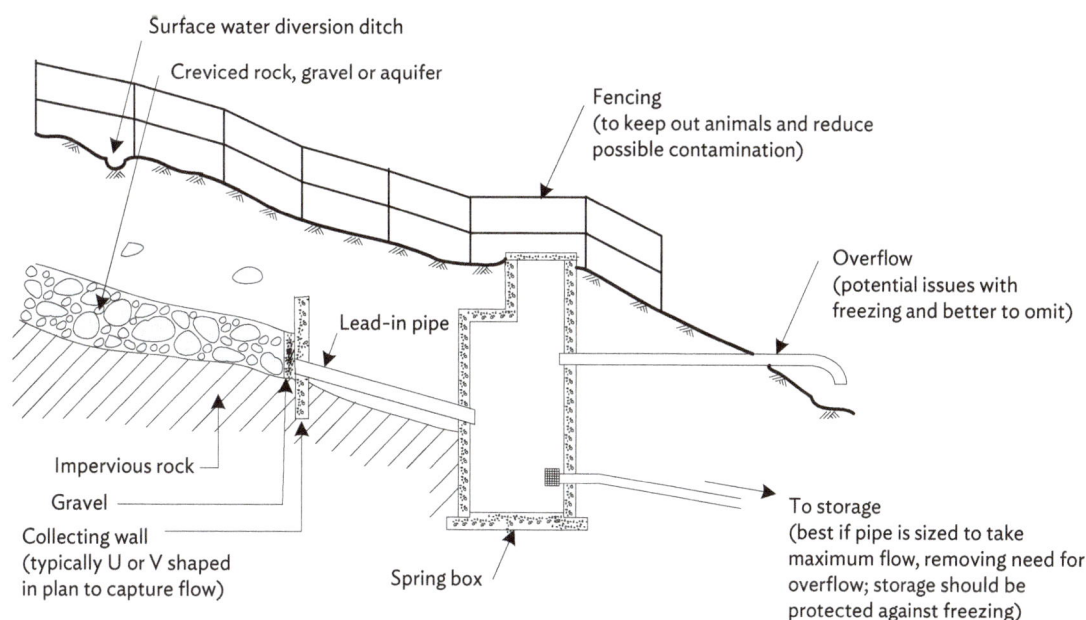

Source: Authors.

The arrangement shown in Figure A3.5 requires modification to be suitable for use in Mongolia. The storage tank should be located in a building and suitably insulated, and the pipe connecting the spring box to the storage tank should be buried to below the level to which the ground freezes. The overflow should be from the storage tank rather than the spring box since water in the pipe from the spring box is likely to freeze. The expense and difficulty of adopting these measures depend on the local topography.

3. Bore Holes and Bore Hole Pumps

In permafrost areas, borewells must extend beneath the permafrost to reach unfrozen water. Permanently unfrozen water may be present below water bodies and perennial streams.[1]

a. Shallow Groundwater

Groundwater located within about 8 m of the surface can be accessed using a positive displacement piston pump located at ground level. Although such systems exist in Mongolia, most piston pumps are hand-operated and are not suitable for use with a piped distribution system. Installation of a hand pump should only be considered as a short-term option for improving an existing open well that is currently accessed using a bucket. In the long term, a piped system should be planned and implemented.

When the pump is not in use, any water left in the upper part of the pump or riser pipe freezes in winter. Both the cylinder and the riser must drain after use. This requires a "leaky" lower valve to the cylinder and a small hole in the riser pipe to be just above or just below the water level (Buttle and Smith 2004 has further information).

[1]   Alter (1969).

Moreover, the draining of the riser between uses may create a need to prime the pump, by adding water to the riser before use, which also increases the difficulties associated with pump operation.

b. Deep Groundwater

When the water table is more than 8 m below the surface and a suction pump located at ground level cannot lift water, a submersible pump located inside the borehole and below the water table will be required. The pump draws in water through a perforated screen and discharges it at pressure into a riser pipe, which connects with pipework that conveys the pumped water either to a storage tank or reservoir or directly into supply.

Box A3.5: Introduction to Deep Borehole Pumps

Centrifugal Pumps

Centrifugal pumps designed for use in boreholes include volute pumps and turbine pumps. Both incorporate a rapidly rotating impeller within a close-fitting case, which draws in liquid at the center and throws it out through an opening or openings at the sides of the casing.

In volute pumps, the clearance between the impeller and the casing increases to a maximum at the outlet. As water flows through the expanding space created by this arrangement, its velocity decreases and the velocity head is converted to pressure head, which lifts the water to the design height.

In a turbine pump, the impeller is surrounded by diffuser vanes, which, like the impeller vanes, are curved and shaped so that the distance between them increases as the distance from the centerline of the pump increases. The conversion of velocity energy into pressure energy takes place mainly between the diffuser vanes.

Volute pumps are most effective for medium to high capacity, and low to medium head applications. Turbine pumps are better adapted to high lifts and seasonal fluctuations in the water level in the well, but are more difficult to install and repair than volute pumps.

Progressive Cavity Pumps

Also known as eccentric screw pumps, progressive cavity pumps are positive displacement pumps, which, unlike piston pumps, do not require valves. The pump consists of a single helix rotor that revolves eccentrically inside a double helix stator, formed by molding the internal surface of the casing to the required double helix shape. This arrangement creates a set of fixed-size cavities between the rotor and the casing. They move through the casing when the rotor is turned, without changing their shape or volume. The water contained in each moving cavity moves with the cavity. As water moves up, it creates a vacuum at the lower end of the pump, which draws in water.

The pressure of the water emerging from the top of the pump into the riser pipe pushes the water already in the pipe upward. Progressive cavity pumps are well suited to pumping relatively small volumes of water against high heads. They require a long drive shaft, which is held in position by water-lubricated "spider bearings" usually made from rubber. Their efficiency is similar to that of centrifugal pumps.

Progressive cavity pumps are a good option for solar-powered installations because they are of variable speed, with the volume of water delivered depending on the speed. This means that they can deliver water across the wide range of power inputs generated by solar power during the day.

The biggest challenge with progressive cavity pumps is to link the centrally located drive shaft to the rotor, which moves in an eccentric motion around the vertical centerline of the pump installation. This linkage is normally provided through a set of universal joints, which transfer power from the concentric rotation of the drive shaft to the eccentrically rotating rotor.

Source: Written by the authors based on a number of sources.

Deep borehole pump options include centrifugal pumps and progressive cavity pumps. Submersible centrifugal pumps are cheaper than positive displacement pumps with the same capacity. The main advantage of positive displacement pumps is that they can pump water over a wide range of power inputs. This means that they are a good option for use with solar and wind-powered systems. Box A3.5 gives further information on centrifugal and progressive cavity pumps.

Most borehole installations connect the riser pipe and the water supply system at the top of the borehole, while making provision for the connection to be removed to allow lifting of the riser pipe and submersible pump for maintenance and repair. This arrangement, however, is problematic where the ground freezes during the winter, because the riser pipe passing through the frozen ground contains water at all times, which freezes up if it is static for any length of time.

Figure A3.6 illustrates two options for dealing with this problem:

(i)   locate the borehole in a heated and well-insulated building; and

(ii)   incorporate a pitless adaptor or pitless unit, designed to allow water to be diverted from the borehole below the frozen ground.[2]

In the conventional arrangement, the riser pipe extends above ground level and a bend leads to the various valves and fittings required at the head of the borewell. In Mongolia's climate, these must all be located within the heated building, and the delivery pipe must either be heated or immediately taken back down to below the frost line, as shown on the left-hand side of Figure A3.6.

Figure A3.6: Arrangements for Conventional Borewell and Pitless Adaptor Installations

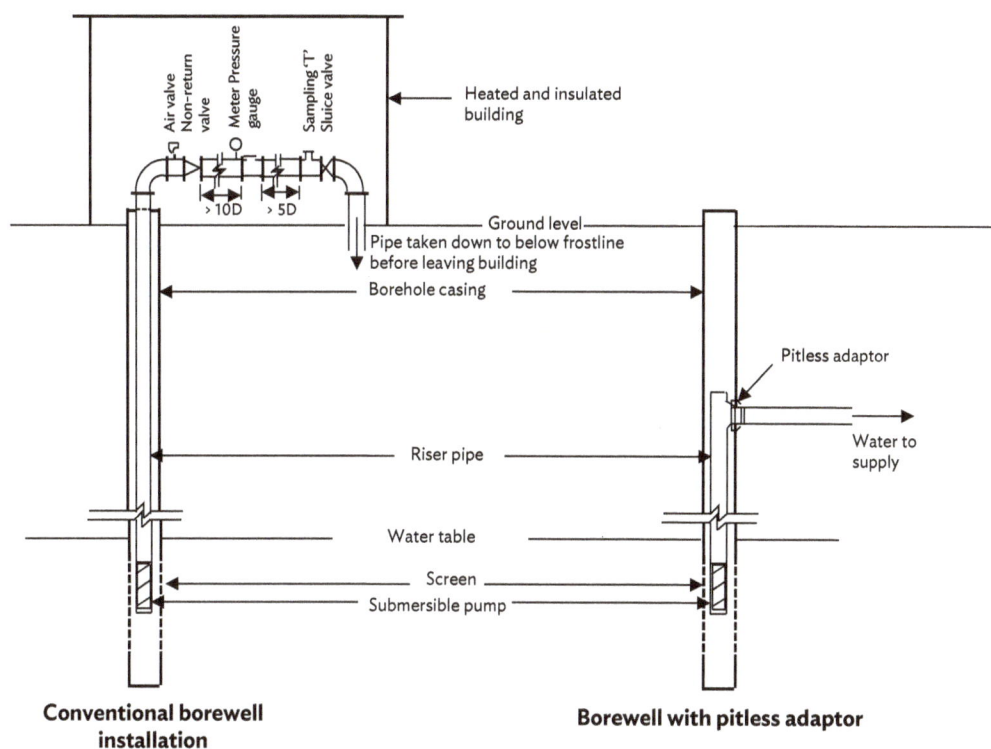

Conventional borewell installation

Borewell with pitless adaptor

Source: Authors.

---

2   Further information on options and installation procedures for pitless units and pitless adaptors is in Alberta Agriculture and Forestry (2012).

A pitless adaptor is a special two-part fitting. The outer part of the adaptor is attached to the inner face of the borewell casing so that it fits around a hole drilled into the side of the casing. The attachment may be made by either using compression gaskets held in place by bolts or clamps, or by welding the attachment to the borewell casing. The horizontal delivery pipe is attached to the outer side of the borewell casing in a similar way. The inner part of the adaptor includes a 90-degree bend and is designed to connect with the top of the riser pipe and into the outer part of the adaptor, thus conveying water from the riser pipe to the horizontal delivery pipe. The adaptor is designed to be watertight. The inner part can easily be disconnected from the outer part so that the whole adaptor, together with the riser pipe and the pump, can be lifted out of the well. When lifting is required, an extension pipe is screwed into a connection on top of the pitless adaptor and the whole assembly is then lifted together.[3]

Pitless units differ from the pitless adaptors in that they replace the well casing above the horizontal feed take-off point. The pitless unit must be manufactured as a complete unit. While this increases the cost of the adaptor, it simplifies construction and may be a better option for Mongolia, where the pitless adaptors or units must be imported.

The discharge piping should incorporate a nonreturn valve, an isolation valve, pressure gauge, totalizing flow meter (a flow meter that records cumulative flow), and a sampling spigot. An air-valve should be included if there is a high point in the pipework, as there is with the conventional pipework arrangement (see left-hand side of Figure A3.5). When the borewell incorporates a pitless unit or adaptor, valves, pressure gauges and meters can be located at a building remote from the head of the borewell. That will particularly be advantageous if supply is from more than one borewell and/or the borewell is at some distance from the school.

Unless a pitless adaptor is fitted, the system must be designed and operated in a way that the water in the pump riser and supply main never freezes. This requires some form of heating in the pump house (photo).

The simplest heating options are solid fuel stoves and boilers but these options involve high operating expenses. Masonry heaters store heat generated by burning wood in a large mass of masonry, located in the building to be heated. The advantage of this arrangement is that the heat is retained in the masonry and released slowly so that maximum room temperatures are reduced while heat is retained for much longer than the case with conventional heaters. Although they are more expensive to install than conventional heaters, masonry heaters appear to be well suited to the task of maintaining the temperature in a building, such as a pump house, above freezing.[4]

Pump house in Tonkhil Soum, Govi-Altai Aimag. *Heating arrangements should be made in the pump house so that the water in the pump riser and supply main never freezes* (photo by Kevin Tayler).

---

[3]   Further information on standards for pitless units and adaptors can be found in Water Systems Council (2013).

[4]   Masonry Heater Association of North America (http://www.mha-net.org/what-is-a-masonry-heater/) has a general introduction to masonry heaters.

Other possibly viable space heating options include solar heating and the generation of solar electricity, which is in turn used to power electrically powered heating systems. Many of these options are at an early stage of development and should only be used at a pilot scale after careful investigation of their potential limitations, strengths, and weaknesses.[5]

Another possible option for reducing the cost of the heating required to avoid freezing is to use the concept of passive solar heating. Over the country as a whole, Mongolia averages more than 250 days of sunshine every year and the possibility of using solar radiation to heat south-facing walls and roofs exists even in the winter. Investigations in Alaska suggest that the most effective strategy for utilization of passive solar heating is to combine south-facing glazing with a thermally efficient building envelope.[6]

Regardless of the type of heater used, heating costs can be reduced if the pump house is well insulated. Box A3.6 provides some basic information on insulation. Standard designs should be developed by an architect with specialist knowledge of insulation techniques and details.

Box A3.6: Insulation Basics

The main heat loss from buildings results from air leakage and conduction. Minimizing air leakage, therefore, is important in extremely cold climates. This can normally be achieved by wrapping the building in a fully adhered membrane with all the insulation outside the membrane and all the services running inside the membrane. At the top, this membrane should be watertight but not vapor-tight. Inside, walls must be vapor-tight and airtight. At the outer part, walls must be watertight but allow vapor to escape so that moisture does not accumulate in the walls and freeze. This is achieved by providing well-engineered cladding.

Layers of rigid insulation should be installed between the interior membrane and the exterior cladding or roof covering, to a minimum thickness of 200 millimeters (mm) on walls and 250 mm on the roof. Special care is required to ensure that air leakage does not occur around windows and doors. Vapor-tight walls prevent vapor from escaping from the inside of a building, leading to condensation problems unless an effective ventilation system is installed.

Conduction occurs where a building component bridges the insulation, resulting in local reduction in the "R" value of the wall. Timber framing has a lower R value than insulation. Therefore, a timber-framed building is less well insulated than the case from the R value of the insulation alone. Unless well-insulated, electrical boxes and other fittings can reduce the thickness of insulation and thus create bridging points for heat conduction.

Experimental work has been done in Mongolia on the use of straw bale walls with a high insulation value. The challenge with straw bale construction, as with other forms of construction, is to prevent moisture accumulation, which leads to rapid deterioration in the condition of the wall.

In permafrost areas, the building must be raised above the ground with a ventilated air space between the building and the ground to control heat lost from the building to the ground. The challenge for a pump house is how to deal with the pipe from the borewell, which creates a bridge between the ground and the pump house.

Source: Cold Climate Housing Research Centre. http://www.cchrc.org/.

---

[5]   The Cold Climate Research Centre provides a general introduction to heating options.
[6]   Further information can be found in the University of Alaska Fairbanks Cooperative Extension Service Passive Solar Heating: An Energy Factsheet (https://www.uaf.edu/files/ces/publications-db/catalog/eeh/EEM-01258.pdf).

Assessing Water Treatment Options

1. Treatment Requirements

If the school has its own supply or if a new supply to a *soum* center includes supply to a school, the need for treatment must be assessed. Deep groundwater is unlikely to contain pathogens and if it is uncontaminated by either chemicals or minerals, it is normally the first choice. All other sources need treatment, varying from disinfection to full treatment including sedimentation and filtration followed by disinfection.

Figure A3.7 illustrates the relationship between water quality and treatment needs.

Figure A3.7: Relationship between Water Quality and Treatment Requirements

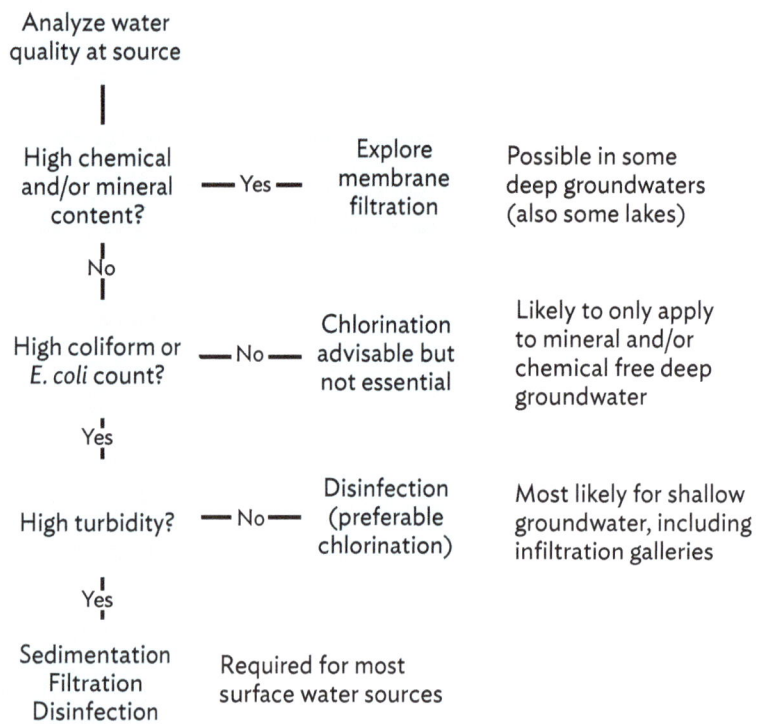

Analyze water
quality at source

High chemical
and/or mineral   — Yes —   Explore   Possible in some
content?   membrane   deep groundwaters
                              filtration   (also some lakes)

No

High coliform or   — No —   Chlorination   Likely to only apply
E. coli count?   advisable but to mineral and/or
                              not essential chemical free deep
                                             groundwater

Yes

High turbidity?   — No —   Disinfection   Most likely for shallow
                              (preferable   groundwater, including
                              chlorination) infiltration galleries

Yes

Sedimentation   Required for most
Filtration   surface water sources
Disinfection

Source: Authors.

2. Treatment Location

Following are the options for treatment location:

(i)   **Prior to distribution.** This is the normal approach to treatment for public water supply systems. Treatment may take place close to production facilities or after transmission. Treatment prior to distribution is normally the cheapest option, in terms of both installation and operational cost, for large public systems. However, it may not necessarily be the case for systems serving schools and/or small settlements. Management may also be an issue where the scheme serves only the school and requires something other than basic disinfection.

(ii) **At the "point of entry".** "Point of entry" is normally used to refer to treatment intended for all the water entering a domestic dwelling. For schools, it is more appropriate to use the term for a device that treats the water delivered to a facility such as a sanitation block.

(iii) **At the "point of delivery".** "Point of delivery" devices are linked to individual water-using devices such as water faucets, sinks, and washbasins. They treat only the water intended for direct consumption (drinking and cooking) and offer the possibility of devolving operational responsibility to those who use a particular water outlet. The linking of responsibility and use should make it easier to ensure good system management.

Industrialized countries such as the United States use "point of entry" and "point of delivery" treatment mainly to remove minerals and chemicals such as arsenic and fluoride. In countries such as Mongolia, removal of pathogens may also be necessary at the point of entry or point of delivery.

For schemes supplied from a public piped system, including new schemes to supply all the public buildings in a *soum* or *bagh*, treatment prior to distribution will be the responsibility of local government.

If the only source for a scheme to serve a single school is contaminated in some way, the advantages and disadvantages of the various treatment options must be assessed.

For a system serving only the school, "point of entry" units may be less costly and easier to install and maintain than a treatment plant to serve all the school's water needs (US EPA 1994). The challenge with "point of entry" and "point of delivery" units, as well as with dedicated treatment plants serving an entire school, is to monitor performance and ensure reliable plant performance. The minimum requirement of any treatment process at the school level is to control microbiological contamination. Conventional systems achieve this through disinfection, while options based on low pressure microfiltration are also possible. For spring and surface water sources, it may be necessary to remove suspended solids prior to disinfection.

3. Treatment to Remove Chemicals and Minerals

Treating water with unacceptably high levels of arsenic, fluoride, and salinity is difficult, both prior to distribution and at point of delivery. Therefore, water from a source with an unacceptably high concentration of chemicals or minerals should only be considered if no other source is available. Where there is no alternative, it is necessary to seek specialist advice on treatment options.[7] Options for "point of delivery" removal of arsenic, chemicals, and minerals are given in Box A3.7.

---

[7]   Further information focusing on arsenic removal can be found in the Sustainable Sanitation and Water Management Toolbox.

Box A3.7: Options for Removal of Arsenic and Other Harmful Minerals

Coagulation and filtration systems. Examples include the two-bucket treatment system and the Stevens Institute system, both of which rely on two containers. In the first bucket or container, a coagulant (materials used include aluminum sulfate, potassium permanganate, iron sulfate, and calcium hypochlorite) is added to the raw water. Settled water then passes into the second container, which is partly filled with a suitable filter media.

Sorptive filtration. This method uses materials such as activated alumina, activated carbon, iron and manganese coated sand, kaolinite clay, hydrated ferric oxide, activated bauxite, titanium oxide, and silicon oxide. Sorptive filtration using materials such as activated carbon also remove fluoride but are less effective in removing pathogens. The effectiveness of the sorptive medium depends on the use of oxidizing agents to promote sorption on the media. Examples of household-level systems based on sorptive filtration include the SONO 3 KALSHI, the KanchanTM, and the SAFI arsenic filter.

Membrane filtration. This includes low-pressure membranes providing microfiltration and ultrafiltration, and high-pressure membranes providing nanofiltration and reverse osmosis. Iron and manganese in water lead to fouling of membranes. Once fouled, membranes cannot be backwashed but have to be removed and cleaned, in accordance with manufacturer's instructions. Membrane filtration also removes pathogens, salts, and various metal ions.

Ion exchange. This can be used in households to soften hard water.

Source: Sustainable Sanitation and Water Management Toolbox. https://sswm.info/sswm-solutions-bop-markets/affordable-wash-services-and-products/affordable-water-supply/arsenic-removal-technologies.

Most of the options set out in Box A3.7 are available as package treatment plants, manufactured and marketed by international companies. Package plants offer a fairly simple option for school water supplies, sited either prior to distribution or, taking a "point of entry" approach, on the mains serving individual sanitation blocks, kitchens, etc. In choosing a package plant, it is important to examine its operational needs, including operational cost, need for chemicals that may have to be brought from a central store, and the need for skilled operators and regular performance monitoring. These needs should be compared with the resources available at the school. If the required resources are not immediately available, options for providing them should be explored.

In removing hardness, nanofiltration is an effective method. Manufacturers produce nanofiltration units for installation at point of use which may be an option if water is too hard to be palatable.

4. Disinfection

For systems served by deep boreholes, disinfection is desirable, while for the water drawn from shallow wells, boreholes, springs, and surface water sources, it is essential. Chlorination is the best disinfection option, since it can create a residual that protects water from subsequent contamination. Other disinfection options include ozonation (treatment with ozone) and ultraviolet radiation. Ozone treatment is a useful option for heavily contaminated water but has few obvious attractions for relatively unpolluted deep groundwater supplies. To date, it has mainly been used at large treatment plants because of its management requirements. It may not be an appropriate choice for school and small community systems.

Ultraviolet (UV) radiation treatment is simpler than either chlorination or ozone treatment and does not require long contact time with the water to be treated. Its main disadvantage is that it leaves no residual to protect the water from subsequent contamination, although this is less of a problem for "point of delivery" and "point of entry" treatment options. Moreover, some pathogens deactivated by UV light may be reactivated when exposed to oxygen. Another disadvantage of UV radiation is that solid particles suspended in water absorb UV light. This reduces its effectiveness as a means of treating heavily contaminated water. These points suggest that UV radiation treatment should only be considered as a disinfection option for clear water.

5. Sedimentation and Filtration

Some form of sedimentation and filtration is usually required for the water taken from a lake, river, or stream. Conventional treatment options fall into two broad categories: those based on slow sand filtration and those based on rapid gravity sand filtration. Since both technologies were developed for use at the settlement level, they are not suitable for school level use without considerable modification and simplification. Researchers have developed a number of household level sand filters over the years that are theoretically suitable for "point of delivery" use. However, because it is difficult to guarantee the quality of water that they produce, their use is not recommended.

Other filtration options at the point of delivery include ceramic filters, activated carbon filters, and the filtration options listed in Box A3.7 on treatment to remove chemicals and minerals.

The flow rate through a ceramic filter is low, typically between 1 and 3 liters per hour for nonturbid water. The effectiveness of ceramic filters in removing bacteria, viruses, and protozoa depends on the production quality of the ceramic filter. While most ceramic filters remove bacteria and the larger protozoans, they are less effective in removing viruses. Studies in developing countries have shown that adequate removal of bacterial pathogens is possible in water filtered through high quality locally produced or imported ceramic filters. For example, a 60%–70% reduction in diarrheal disease incidence has been reported from use of high-quality filters. On the other hand, studies have also revealed significant bacterial contamination when poor-quality, locally produced filters are used, or when the receptacle is contaminated at the household level. Because there is no chlorine residual protection, it is important to train users in the proper care and maintenance of ceramic filters.[8] Regular cleaning of filters is essential, especially when the water to be filtered is turbid.

Assessing Transmission, Distribution, and Storage Options

Wherever possible, the school should be served with a piped water system. If, however, there is no local source and funds to construct a transmission main from a remote source, trucking of water may be necessary. This option is expensive and should be viewed as a temporary measure, to be replaced by a fully piped system as soon as possible.

When planning a water delivery system, it is necessary to consider the following:

(i)   system configuration, including the location of storage;
(ii)   provision for protection against freezing;
(iii)   design flows and pressures;
(iv)   pipe materials and minimum diameters; and
(v)   amount of storage to be provided, as well as storage tank materials.

---

[8]   More information on ceramic filtration can be found in the Centers for Disease Control and Prevention (http://www.cdc.gov/safewater/ceramic-filtration.html).

1. System Configuration

All piped systems must include a distribution system and storage. Distribution mains deliver water from a central point to sanitation blocks, kitchens, laundry rooms, and other places where water is needed. Storage allows for variations in water use (demand) over a typical day and ensures continuity of supply. A transmission main is required if the source is at some distance from the school.

The storage volume required to cater for variations in demand depends on the supply regime (i.e., at what times during the day are the pumps that supply the system operated) and the demand pattern. The demand pattern is roughly the same for most systems, that is, demand is greatest in the morning when people rise, wash, and eat; fluctuating during the day; lowering in the early evening; and dropping to almost nothing overnight. If water is to be available throughout the day, around 6-hour storage will be required at the maximum to meet the average daily demand. In practice, this will normally equate to around 12-hour storage at the average demand throughout the year.

Storage may be provided at ground level or in a tank elevated to a height that allows water to gravitate to taps and other water-using devices. Elevated tanks have the advantage that they allow gravity flow to washbasins, showers, and sinks, eliminating the need for pumping other than at the water source. Conventional practice in temperate climates is to provide a central elevated tank, from which distribution mains convey water to all the locations where water is required. The challenge with this approach in cold climates is to ensure that the water in the system does not freeze.

Figure A3.8 shows a simple system with a transmission main from a borewell to an elevated storage tank, from which a branched distribution system conveys water to the sanitation blocks and kitchen area. The arrangement assumes that the borewell pump delivers water directly to the elevated storage tank. Gravity supply may be possible in rare situations where supply is from a surface source or infiltration gallery located at a higher elevation than the school so that water can flow to the school by gravity.

Figure A3.8: Branched Distribution System from Central Elevated Storage Tank

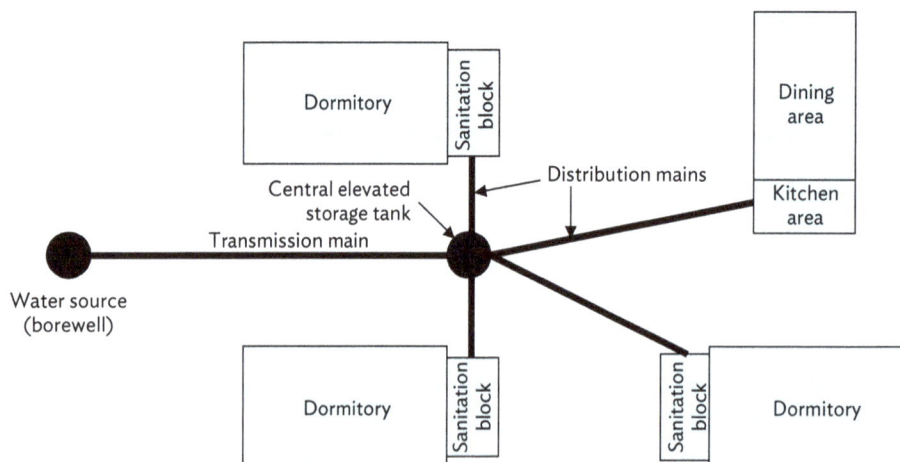

Source: Authors.

Storage could be provided in several smaller tanks, located in or close to the buildings to be supplied, rather than centrally. Distribution could be via a looped system rather than a branched system. Figure A3.9 shows a system incorporating these options configured to serve the same buildings as those shown in Figure A3.8.

Figure A3.9: Looped System with Decentralized Storage

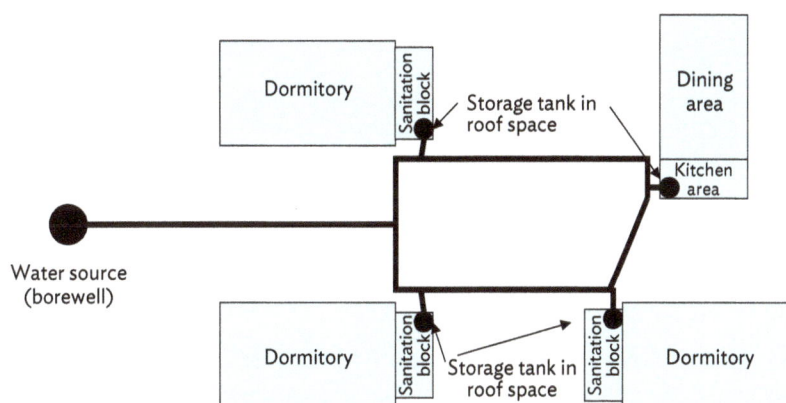

Source: Authors.

Whenever possible, it is better to provide a looped system, especially in cold climates. The looped system improves water circulation and reduces the risk of water becoming stagnant. Since stagnant water freezes before moving water, the looped system is slightly less susceptible to freezing than the branched system.

Decentralized storage has two potential advantages. First, several small, prefabricated tanks may be cheaper to install than one large tank built *in situ*. Second, there is a possibility of siting smaller tanks in the roof space of buildings or on upper floors of buildings, where heating for the buildings can provide protection against freezing.

The challenge with decentralized elevated storage is to find space to locate storage tanks within buildings. For buildings with pitched roofs, the tank can be located within the roof space, as shown in Figure A3.10. The tank should be fitted with a floating ball valve to close the inlet when the tank is full to prevent overflowing.

Figure A3.10: Water Tank Located within Pitched Roof

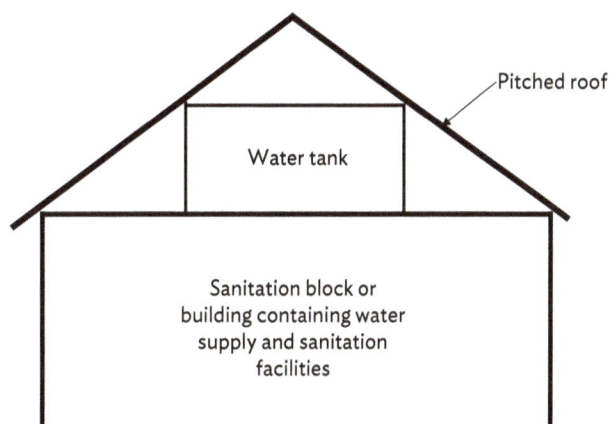

Source: Authors.

If a new building is constructed to house water supply and sanitation facilities, space should be provided to allow a water storage tank to be placed within the insulated building space. Additional insulation should be placed around the tank to ensure that its contents do not freeze.

Careful planning to minimize the length of water mains would bring greater benefits in cold than in temperate and warm climates because the cost of construction is higher in cold climates.

2. Protection against Freezing

Water mains buried at a "normal" depth of around 1.25 m, as well as external storage tanks, freeze in Mongolia's harsh winter climate. Without effective measures to prevent freezing, the water supply system will quickly fail. Measures to prevent freezing include burying the main below the lowest depth to which the ground is frozen or heating the main in some way.

Below are options for preventing freezing with their advantages, disadvantages, and limitations.

(i)   **Bury pipes below the depth at which the ground freezes.** The construction cost of this option will be high, while operational costs will be low. One potential problem would be the cost and difficulty associated with repairing pipes. Using long lengths of polyethylene pipe reduces the potential for leaks and the need for pipe repairs. The approach, however, is not possible in areas that are subject to permafrost.

(ii)   **Heat pipes.** Pipes can be heated by using an electrically powered heat trace or laying pipes inside or alongside a heated conduit (known as a utilidor). The heat trace option will have a low capital cost but would be expensive to run. The utilidor option is only viable in those areas of Ulaanbaatar, *aimag* and *soum* centers where district heating is provided or planned, and in schools that have their own centralized heating systems.

(iii)   **Circulate heated water through a looped system.** This approach requires the circulated water to pass through a heat exchanger. Drawbacks are its relative complexity and its high operating cost.

All three heating options would be expensive to operate. Thus, the only realistic option for schools located in areas without existing district heating is to bury the pipe deep enough to prevent freezing. The required depth varies across the country but is normally 2.5 to 2.7 m. Electric heat tapes might be used to protect pipes where they rise to the surface. It might also be possible to use photovoltaic panels to generate the electricity needed to heat them, although this option needs storage batteries. Where heat tapes pass along or through insulation material, the material should be nonflammable such as fibrous glass.

Supply pipes should be routed to within buildings wherever possible, preferably on internal walls. Pipes on external walls should be located within any insulation. Internal pipework should be sloped slightly and a tee with a valve on the branch should be provided at a low point to allow pipes to be drained when necessary.

Insulation reduces heat exchange between the water contained in a pipe or storage tank and the ground or air that surrounds the pipe or tank, thereby increasing the time taken for the contents of a pipe or tank to freeze. For countries such as Mongolia that are subject to several months of subzero temperatures, the water in pipes and tanks freezes eventually unless it is either regularly replaced by warmer water or heated in some way. Insulation, therefore, is most effective if combined with pipe heating or circulation of water within the distribution system.

Mains must pass through the freezing zone at the point where they connect to buildings. The heat that the building generates provides some protection against freezing, particularly where the pipe rises to the surface directly below the building. However, there are many situations in which pipes need to be insulated to minimize the risk of freezing. Polyurethane and polyethylene foam are good insulators. Other options include glass and mineral wool. A typical arrangement is to surround the water main in a suitable insulating material and to enclose both in a larger diameter outer pipe (Figure A3.11).

Figure A3.11: Typical Insulation Arrangements for Water Pipe Passing through Freezing Zone

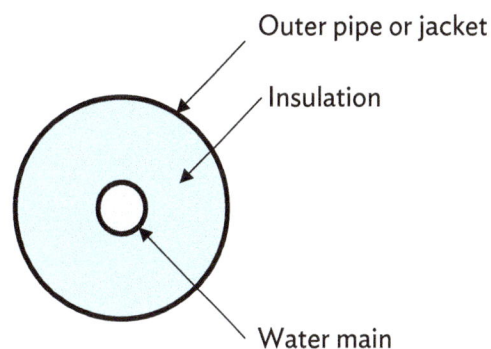

Source: Authors.

Protection against freezing is also required for water storage tanks. As already indicated, a good option is to provide a storage tank for each sanitation block or kitchen area and to site it inside the building so that the tank can benefit from the heating provided for the building. If a new building containing the facility is planned, the design should include provision for installing an elevated tank. If improved sanitation, kitchen, or laundry facilities are retrofitted into an existing building, provision of elevated storage within the building may be more difficult.

When options to provide protection against freezing for a storage tank are considered, two questions should be asked: Where can the storage tank be placed? And how can it be supported? For buildings with pitched roofs, the answer to the first question might be "in the roof space." The response to the second question usually involves structural strengthening, which requires input from a structural engineer.

Where supply is via a central elevated tank, the tank should be either buried at least 2.5 m beneath the ground at an elevated location close to the school, or well insulated and heated. One option, albeit expensive, may be to house the tank in a building. When exploring this option, the possibility of using the space below the tank as a store should be considered. Regardless of where the tank or tanks are located, insulation is required. Insulation materials for tanks include polyurethane and polystyrene sheet.

3. Design Flows and Pressures

Water mains must be designed to carry the maximum predicted flow at the design horizon, typically 20 to 25 years. Calculation of the maximum flow for a branch main supplying a dormitory sanitation block is given in equation (3):

$$\text{Flow (l/sec)} = N \times q \times pf /(3600*24) + \text{allowance for leakage and wastage} \qquad (3)$$

where

> N is the number of students using the block,
>
> q is the per capita water consumption, and
>
> pf is the relevant peak factor.

The peak factor used depends on whether the main is delivering directly to taps or to a storage tank. The peak hour factor should be used for the first case and the peak day factor for the second case. For looped systems, flows, pipe sizes, and pressures can be determined using a simple computer program such as EPANET, which can be downloaded free of charge from the internet.[9]

The pressure at taps depends on the vertical distance between the taps and the minimum water level in the elevated tank serving them, which should not be less than 3 m. Similarly, the supply system from the water source or sources to elevated tanks should maintain a minimum head of at least 3 m above the top water level of the highest elevated tank.

4. Pipe Material and Diameter

Materials commonly used for water distribution pipes include ductile iron, unplasticized polyvinyl chloride (uPVC), high and medium density polyethylene, and cross-linked polyethylene. Copper is the most commonly used material for internal pipework. Ductile iron, which is expensive, is only an option for larger diameter pipes, and is unlikely to be used for school water supply systems. Box A3.8 summarizes the advantages and disadvantages of these options.

Main diameters depend on the calculated flow. External mains should not be less than 50 mm in internal diameter. Recommended minimum diameters for internal pipework are as follows:

Pipe serving a single washbasin or sink	20 mm
Pipe serving a single toilet cistern	12 mm
Pipe serving a row of washbasins	25 mm
Pipe serving a row of toilet cisterns	20 mm
Main from internal storage tank	38 mm

---

9   US EPA. EPANET: Application for Modeling Drinking Water Distribution System. http://www.epa.gov/water-research/epanet.

Box A3.8: Pipe Material Options for School Water Supply Systems

uPVC pipes. They are widely used and comparable in cost to other plastic pipes, but have some disadvantages compared with other plastic pipe options. They normally come in 6-meter (m) lengths with a push-fit jointing arrangement that incorporates an O ring. Jointing needs care as a badly jointed pipe would leak. The rigid nature of uPVC means that a pipe shatters if the water inside freezes.

Polyethylene. This elastic material can expand to some extent under pressure and then return to its original size, making it more resistant than uPVC to the effects of freezing. Tests suggest that the elasticity of polyethylene decreases when it is repeatedly subjected to a high level of stress. Therefore, the long-term integrity of polyethylene pipes, when exposed to freezing conditions, cannot be guaranteed. Polyethylene pipes with diameters of 50 millimeters and less come in long lengths coiled onto reels. This significantly reduces the number of joints required, which in turn reduces the likelihood of leakage from joints. Larger diameter pipes can also be supplied on reels but normal practice is to provide them in lengths of around 15 m. Options for joining pipes include electrofusion and compression fittings.

Cross-linked polyethylene. Also called PEX, this material is commonly used for internal pipework. Like polyethylene pipes, PEX pipes are elastic, which means they resist freezing better than more rigid pipes. However, like polyethylene pipes, they would fail if subjected to repeated freezing and thawing cycles. Cross-linked polyethylene pipes should not be used or stored where they might be exposed to sunlight, as it can degrade the pipe material.

Copper pipes. They are the more "traditional" choice for internal pipework. Joints are soldered or made using push-connect fittings. Since they are less elastic than cross-linked polyethylene pipes, copper pipes are more susceptible to damage if the water inside freezes.

Source: Compiled by the authors from various sources.

5. Storage

The storage required depends on the supply regime. From the demand pattern over a typical day, it is possible to calculate the volume of storage required. For systems with a reasonably constant supply, a tank that provides 12 hours storage at the average daily demand should be more than adequate to deal with normal fluctuations. If the supply regime is irregular, more storage may be required. Specialist advice can be sought on calculating the required storage capacity, if necessary.

Larger tanks are usually constructed using concrete or modular steel panels. For capacities up to about 10 cubic meters, prefabricated plastic tanks are available, mostly fabricated from polyethylene. Each storage tank should be fitted with a floating ball valve that closes when the water level in the tank approaches overflow level. Overflow pipework should run within buildings to a discharge point located just outside the building. This ensures that a trickle of overflow water does not freeze and block the pipe. The discharge point should be at a point that is visible and accessible so that overflow from the tank can be observed and appropriate action taken.

Figure A3.12 shows a typical high-level storage tank, intended for installation within a building. As already indicated, the challenge in existing school buildings is to find space and to provide the structural support required for such a tank. The design of new buildings should always include consideration of the options for providing elevated storage. Where installation of an elevated storage tank is not possible, the tank should be installed below ground level with a small pump to pressurize the water delivery system.[10]

---

[10]   Seifert (2004) provides information on storage tank options for Alaska.

Figure A3.12: General Layout of Typical Elevated Storage

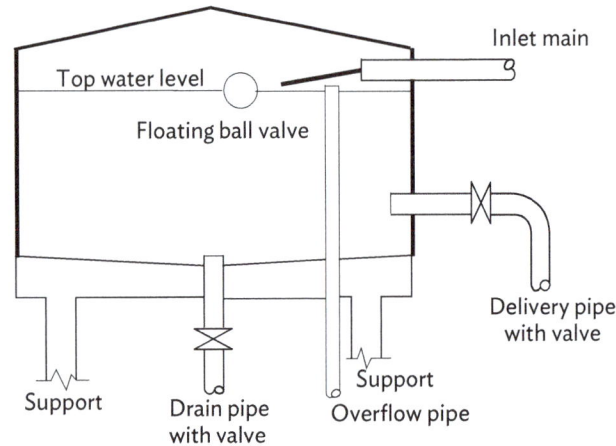

Source: Authors.

Assessing User Interface Options

The last link in the chain connecting the water source to the user is provided by the taps and other water-using devices (basins, sinks, etc.) that enable users to access the water. Water used for washing should be heated in some way to increase the likelihood that students follow proper hygiene practices. Options for designing user interface facilities in a way that users have unimpeded access to water, while minimizing water wastage, are examined below.

To ensure reliable operation, taps should be of a robust design, available in the local market, and of a type that local plumbers are familiar with. In theory, taps that deliver a preset volume of water, once they have been pressed, save water. However, experience suggests that they are more likely to fail than simpler taps, in part because local plumbers often do not have the spare parts to repair them, particularly in remote *soums* and *baghs*. Although it is worthwhile to investigate the possibility of using water-saving taps, they should not be specified unless systems to repair and replace them exist.

If toilets are fitted with cisterns, those cisterns should be fitted with a float valve to prevent overflowing. Some cisterns flush using siphonic action, while others rely on a valve in the bottom of the cistern. Although the valve-type cistern is more efficient in theory, cisterns fitted with bottom valves tend to leak, which can be a serious disadvantage when water availability is limited. If this is the case, it may be better to rely on pour-flush toilets. A tap should be available close to toilets so that the buckets or containers used to flush the toilets can be filled.

There is scope to use solar thermal systems to heat water for delivery via hot taps in washrooms, given that Mongolia has around 250 sunny days each year and sufficient daylight hours varying from a minimum of around 8 hours in December to 16 hours in June. Two types of solar heating, flat plate collectors and evacuated tube collectors, are available.

A flat plate collector consists of a shallow insulated metal box with a clear upper face to let sunlight in and a dark absorptive plate at the bottom, which heats up when exposed to the sun's rays. Heat from this plate is transferred to a mixture of glycol and water, which circulates through copper tubing attached to the absorber plate, then through a continuous run of pipework to a heat exchanger fitted into a water storage tank. This transfers heat to the water in the tank, thereby raising its temperature.

An evacuated tube collector has a series of evacuated glass tubes, each of which contains a smaller copper tube (or tubes) filled with fluid, with vacuum space in between. The purpose of the vacuum is to increase efficiency by reducing heat losses due to convection and conduction. The sun's rays heat the fluid inside the copper tubes to boiling point where its heat is transferred to a glycol mixture, which circulates and heats water in a storage tank, as is the case with the flat plate collector. Once it has lost its heat in the heat exchange manifold, the vapor inside the tube returns to a liquid state and runs back down the tube where it is heated again, repeating the cycle.

Both systems can work in cold climates, provided that there is sufficient glycol in the glycol-water mixture, which should not freeze in winter. While flat plate collectors are less efficient, they shed snow better than evacuated tube collectors. Given Mongolia's low rainfall during winter, the latter point is unlikely to be of major significance. Whichever form of collector is used, it should be erected at an angle that optimizes performance during the winter when the sun is low. It is better to consult with a specialist solar panel provider at the detailed design stage.

APPENDIX 4
Assessing Sanitation Options

Emphasizing the need to consider all links in the sanitation service chain, and resilience of the system under cold winter conditions, Appendix 4 outlines possible responses to poor sanitation conditions in Mongolian schools, and provides guidance on the design of sanitation blocks or washrooms, and the arrangements for the treatment and disposal of excreta and wastewater. The objective is to assess options for meeting government standards and regulations for sanitation facilities, including types and locations of toilets, and the ways in which toilet type influences choices for subsequent links in the sanitation chain. This appendix is intended for professional engineers and architects who are responsible for planning and designing school water, sanitation, and hygiene (WASH) systems.

The term sanitation as used here means the safe management of excreta and wastewater from toilets, washbasins, sinks, showers, and other water-using facilities and activities. Excreta comprise feces and urine, and the combination of feces, urine, and water flushed from a water closet is black water. Gray water is water from bathrooms, sinks, and floor washing. When combined, black and gray water become sewage.

This appendix covers the following:

(i)   state of sanitation in schools in Mongolia;
(ii)   objective of improving sanitation in schools;
(iii)   sanitation service chain;
(iv)   sanitation and wastewater disposal options;
(v)   identification of an appropriate sanitation option;
(vi)   immediate low-cost improvements in school sanitation;
(vii)   washroom and sanitation block design;
(viii) sewerage and drainage design;
(ix)   wastewater treatment and disposal options;
(x)   dry sanitation options; and
(xi)   sludge removal, treatment, and disposal options.

State of Sanitation in Schools in Mongolia

Sanitation in many schools in rural and peri-urban areas currently takes the form of pit latrines, located within school compounds but at some distance from school buildings. Many of these existing pit latrines have a superstructure and floor slab constructed from wood, lack doors, and are in poor condition, with partly collapsed pits and gaping holes on floors. There are no formal systems for removing the contents once the pits are full.

The following photos are show examples of a toilet block in poor condition and a recently constructed block, both in schools in Govi-Altai Aimag.

Pit latrine in school. Note the lack of doors and gaping hole in front of one cubicle (photo by Kevin Tayler).

New pit latrine. Though the toilet is in better condition, there is still no place for children to wash their hands after defecating (photo by Kevin Tayler).

Even when new and in reasonably good condition, outdoor pit latrines are unsatisfactory in several respects due to

(i)   the absence of a tap, which greatly reduces the likelihood that users wash their hands after defecation;

(ii)   the need for users to go out into the cold to defecate, which is especially difficult under Mongolia's harsh winter conditions;

(iii) the lack of privacy; and

(iv)   the absence of lighting, which makes the facilities unattractive and unsafe to use at night.

The poor condition of many school sanitation facilities discourage students from using them. Surveys conducted in *soums* of Khuvsgul Aimag under the UNICEF–AusAID project suggest that many students who live at home avoid using school toilets, and that toilets serving dormitories are underused.

Some schools have water-flushed toilets connected to septic tanks. However, these systems are not always operational, often because the septic tanks and the pipework leading to them are located outside buildings and have suffered frost damage. The photo shows an example of a nonfunctional septic tank.

Nonfunctional septic tank. *Factors contributing to the failure of the septic tank include freezing of the connecting pipes and lack of water* (photo by Kevin Tayler).

Although some sewers exist in *soums*, most have failed due to freezing during cold winters. On the other hand, Ulaanbaatar and *aimag* centers are largely connected to sewers that are laid close to district heating pipes to prevent freezing.

Operation of sewage treatment plants can be problematic. In 2001, out of 104 treatment plants in Mongolia, only 27 were operating satisfactorily while 35 were out of use (Bathold, Tuul, and Oyun 2004). Part of the problem lies in the design of treatment plants, many of which were constructed during the period of a centrally planned economy, require skilled operators, and have high operational costs.

These experiences point to the need to consider operational requirements when assessing options for school sanitation. If treatment plants serving *soums* prove difficult to operate and maintain, there will certainly be problems with similar plants serving individual schools. These problems and possible ways of overcoming them are discussed in the following sections.

Objective of Improving Sanitation at Schools

The overall objective of improving sanitation at schools is to provide sanitation facilities that are safe, convenient, and pleasant to use. Achieving this objective requires that

(i)   toilets are conveniently located,

(ii)   there are sufficient toilets for students to use without having to queue for long periods, and

(iii)   toilets are in a warm and light environment with access to water.

Box A4.1 summarizes the minimum requirements for WASH in schools jointly issued by the Ministry of Education, Culture, Science and Sports (MECSS); the Ministry of Health; and the Ministry of Finance in 2015 (further information is found in section II).

Box A4.1: Requirements for the Location and Number of Sanitation Facilities

Toilets for schools and kindergartens must be indoors.

Dry on-site sanitation facilities should be located at least 20 meters (m) from a dormitory block or communal building (including classrooms), 25 m from water distribution kiosks and deep water wells, and at least 200 m from other water sources, including rivers and streams.

Outdoor toilets should be accessible in all weather conditions and students should not have to walk more than 30 m to an outdoor toilet. They must have insulation, ventilation, and lighting; must be lockable from the inside; and be equipped with a waste bin with lid.

At least one toilet should be provided for every 30 girls, with at least one separate toilet for female staff. For kindergartens, there should be one toilet for every 15 children.

At least one toilet and one urinal should be provided for every 40 boys, with at least one separate toilet and urinal for male staff.

At least one handwashing point should be provided for every 40 students in schools and for 30 students in dormitories. For kindergartens, there should be at least one handwashing point per 20 children.

Source: Ministry of Education, Culture, Science and Sports; Ministry of Health; and Ministry of Finance (2015).

Sanitation improvements will only lead to health benefits if they are accompanied by improvements in personal hygiene, in particular, with regular handwashing after defecation. Therefore, an integrated approach to the provision of toilets and handwashing facilities should be adopted in the design. That is, facilities for handwashing with soap must be located close to toilets. In cold climates like Mongolia, warm water should also be available for handwashing.

Sanitation Service Chain

Sanitation is not only about toilets, but rather, the complete service chain from the toilet to the final safe disposal or reuse of excreta. The Water and Sanitation Program identifies five links in the chain: containment, removal, transport, treatment, and reuse or disposal. The sanitation value chain of the Bill & Melinda Gates Foundation includes five, yet different links: capture, storage, transport, treatment, and reuse. The term "sanitation value chain" used by the Bill & Melinda Gates Foundation highlights the belief that excreta are a potential resource and should not be viewed solely as a problem.

Combining the two chains gives a chain with the following six links:

Capture–Storage–Removal–Transport–Treatment–Reuse or Safe Disposal

Options for capturing excreta range from a simple hole in a slab through pour-flush and cistern-flush toilets, to urine-diversion toilets designed to separate feces from urine. Storage is only required for systems that retain solids on-site. Removal and transport involve either periodic emptying of pits and tanks or connections to sewers to carry excreta flushed from toilets off-site. Systems that retain fecal solids on-site normally require separate drainage facilities to carry storm and sullage or gray water off-site. On the other hand, all systems require some form of treatment, of wastewater in the case of sewered systems, and of septage or fecal sludge in the case of on-site systems. Additionally, provision for using treated sludge and/or wastewater as a soil conditioner or fertilizer can be included. This completes the "closed loop" cycle, which returns nutrients and carbon to the soil rather than allowing them to wash away into the sea or a saline lake.

Sanitation and Wastewater Disposal Options

Sanitation options are interrelated. In particular, the type of toilet influences the range of possibilities for the location of toilet, and subsequent links in the sanitation chain. Choices to be made relate to:

(i)   type of toilet;
(ii) location of toilet; and
(iii) subsequent links in the sanitation chain, i.e., arrangements for storage, removal, transport, treatment, and either safe disposal or reuse of excreta.

1. Type of Toilet

There are two basic types of toilets:

(i)   "wet" toilets incorporating a water seal and requiring water for flushing; and
(ii) "dry" direct-drop toilets.

Wet systems separate users from excreta and reduce the possibility of fly, cockroach, and odor problems. For these reasons, they are widely regarded as providing a higher level of service than dry systems. In practice, however, they only work effectively if sufficient water is available.

Taken together, this suggests two points:

(i)   Water-flushed toilets should be the preferred option for schools with a functioning piped water supply.

(ii)   Some form of dry toilet will usually be the only practical option for schools without a functioning piped water supply.

Toilets can be flushed either from cisterns attached to the toilets (cistern-flush toilets), or by pouring water from a bucket or other suitable container kept near the toilet (pour-flush toilets). In most schools, pour-flush toilets will be the better choice, since they have no mechanical parts that could go wrong, and use less water than cistern-flush toilets.

Water-flushed toilets may be either "western" style, designed for sitting, or "eastern" style, designed for squatting. The choice between the two styles should take account of user preferences.

All dry toilet designs consist of a hole in a slab or floor through which excreta drop into the vault, pit, or container. In urine-diverting dry toilet (UDDT) designs, the hole is incorporated into a prefabricated pan that separates feces from urine, with feces dropping into the pit or vault while urine is piped to a separate container.

The UDDT option has two advantages over conventional dry systems:

(i)   It separates urine, which contains a high proportion of the nutrients present in excreta but virtually no pathogens, from feces, which may contain large numbers of pathogens.[1]

(ii)   It usually smells less than systems that use the same pit for feces and urine.

A challenge for a cold country such as Mongolia is to ensure that urine pipes and containers do not freeze. Some UDDT designs have been piloted in Ulaanbaatar, with the intention of using urine and treated fecal sludge as agricultural inputs. Results, however, have been disappointing and none of the UDDT initiatives has gone to scale.

2. Location and Layout of Toilets

The national minimum requirements for the location of sanitation facilities listed in Box A4.1 indicate that

(i)   water-flushed toilets can be located within existing school buildings or in new extensions to existing buildings; and

(ii)   dry toilets must be located at least 20 m from dormitories and other school buildings.

Although the direct-drop toilets (photos on page 89) comply with the second requirement, they fail to have insulation and lighting and are located at some distance from handwashing points. For schools without a piped water supply, the only feasible option for complying with the national minimum requirements will be to provide waterless toilets in separate sanitation blocks located at least 20 m from other buildings. Therefore, it is necessary to retain direct-drop toilets in the short term. Yet the long-term objective should be to replace them with hygienic toilets grouped in sanitation blocks provided with insulation and lighting.[2] The blocks should include handwashing facilities where water brought by hand is available.

---

[1]   The point is important to note when urine and feces are used as an agricultural input (i.e., fertilizer, soil conditioner).

[2]   There is a strong case for revisiting some aspects of existing regulatory requirements. In particular, it is possible that improved forms of dry sanitation allow its location closer to school buildings than the currently specified 20 m.

Whenever possible, toilets should be located inside or contiguous with an existing or planned building. This will ensure maximum comfort and convenience for users. It will also reduce heat loss through external walls and facilitate extension of existing school heating systems to serve toilets and washrooms, thereby reducing the cost of heating. There is evidence that some detached sanitation blocks suffer from condensation problems and are not used to their full potential because schools were unable to meet the high cost of heating those sanitation blocks.[3]

Toilets, washbasins, and showers for use during the day should be centrally grouped in washrooms or sanitation blocks, located close to classrooms and the area where students eat, so that students can wash their hands before eating.

In addition, washrooms and sanitation blocks providing toilets, washbasins, and showers are required to serve dormitories. The preferred arrangement will be for each dormitory to have its own facility, although the possibility of providing one sanitation block to serve two or more closely spaced dormitories could be considered. The number of each type of unit provided should, at the very least, meet the national minimum requirements. Where a sanitation block is not inside or contiguous with the building that it serves, a lighted and enclosed access corridor is required to facilitate student access to the block under all weather conditions.

Possible locations of sanitation facilities should be assessed in consultation with teachers and school staff, students, and other stakeholders. A plan of the school will be a useful aid when assessing possible locations for washrooms and sanitation blocks (Appendix 2). The plan should preferably be to scale and show the location of all classrooms, dormitories, and other buildings, together with features such as steep slopes and access routes that might influence the location of facilities. Google Earth imagery may provide sufficient information to develop a basic plan. Further details can be added, based on observation and simple ground surveys.

Another approach would be to encourage teachers and students to prepare a sketch showing the location of school buildings, any existing water supply and sanitation facilities, and potential sites for future facilities. The sketch can then be used as a basis for discussion of the possible locations for planned facilities. Teachers and students should be reminded of the national minimum requirements and other relevant standards and regulations. The results of this exercise can be refined with the help of technical specialists.

Points to consider when assessing the location of sanitation facilities are as follows:

(i)   Showers need to be close to sports halls or gymnasiums.

(ii)   Toilets and washing facilities for use in the day should be centrally located in the school, reasonably close to classrooms and any central assembly points such as dining hall.

(iii) Each dormitory should have its own sanitation block, providing toilets, handwashing facilities, and showers. The number of each type of unit to be provided depend on the number of students who are supposed to use the block.

Facilities for boys and girls should have different entrances and be separated by solid walls that reach to roof level. The doors of individual toilet cubicles should be securable and reach down to floor level.

Figures A4.1 and A4.2 show indicative layouts for girls' and boys' toilets intended for general use. [4] Girls' toilets provide four toilet cubicles and three washbasins; the boys' toilets provide three toilets, three urinals, and three

---

[3]   Alcanz Consulting Group (2015) presents the findings of an evaluation of a school WASH improvement program in Khuvsgul Aimag.

[4]   The indicative layouts are taken from Annex 4 of the UNICEF Mongolia report (UNICEF Mongolia 2012).

washbasins. It should be noted that the layouts were developed to fit into converted shipping containers, and the clearances should ideally be slightly greater than shown.

Figure A4.1: Indicative Layout of Girls' Toilets	**Figure A4.2: Indicative Layout of Boys' Toilets**
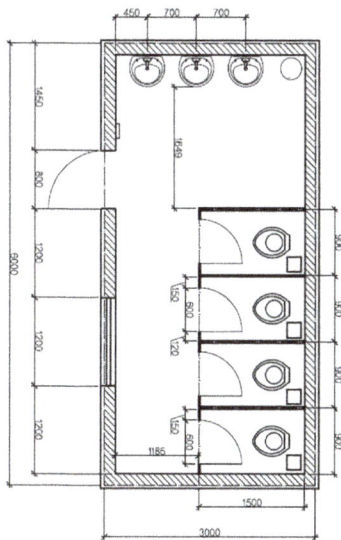	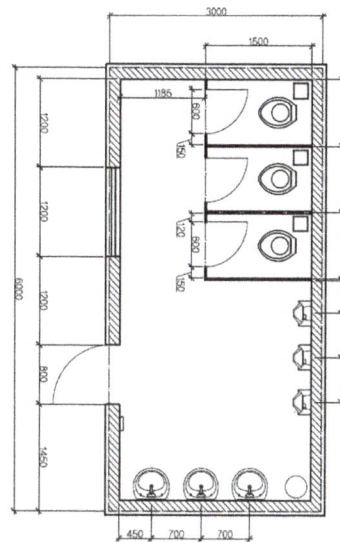
Source: UNICEF Mongolia (2012).	Source: UNICEF Mongolia (2012).

Figure A4.3 give indicative layouts of girls' and boys' toilets for a dormitory. They show (but do not label) a shower cubicle on the left-hand side for the boys' toilet block, and the right-hand side for the girls' toilet block. To meet the national minimum requirements, at least one additional shower cubicle should be provided for both girls' and boys' toilet blocks. In addition, at least one of the toilet cubicles in each block should be larger and equipped for use by students with disabilities. With more space available, the layout of the boys' toilets could be adjusted to remove the direct line of sight from the door to the urinals.

Figure A4.3: Indicative Layouts of Toilets Attached to Dormitory

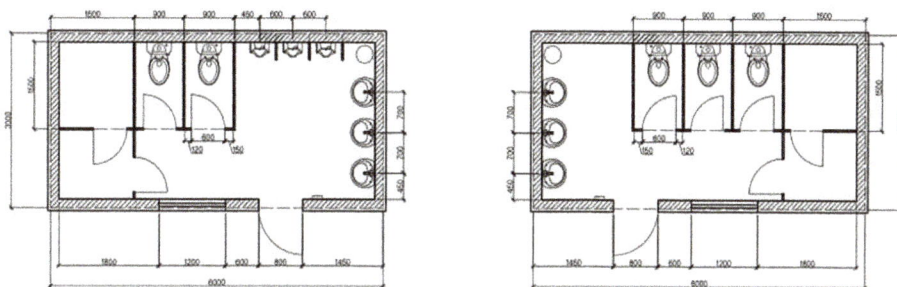

Source: UNICEF Mongolia (2012).

3. Water-Flushed Systems

The type of toilet influences choices regarding subsequent stages in the sanitation service chain. For water-flushed systems, options include:

(i)   water-flushed toilets–sewer connection–sewer–wastewater treatment plant;
(ii)   water-flushed toilets–septic tank–soakaway or drainfield system;
(iii) water-flushed toilets–decentralized treatment plant, located close to the sanitation block–soakaway or drainfield system.

The sewered option is only possible if sufficient water is available to transport fecal solids along the sewer. It should only be considered if per capita water consumption equals or exceeds 60 liters per capita per day (LPCD).[5] If a school is in an area with existing public sewers, the school's sewers should discharge to the public system. If there is no public sewer system but a good water supply, the school's system must include appropriate provision for wastewater treatment, using one of the two options listed above.

If water availability is sufficient to support the use of pour-flush toilets but insufficient for sewerage, one of the other two options must be adopted. The septic tank option is simpler and does not require a reliable power supply or skilled operational staff, both of which are required for the wastewater treatment plant option. For this reason, the septic tank option will normally be the preferred option.

Sewered systems and on-site wastewater disposal systems should be designed to take sullage water from sinks, showers, and washbasins as well as black water from flushed toilets. While off-site systems incorporating sewers can deal with both black water and gray or sullage water, most on-site flushed systems deal only with black water. Therefore, separate provision must be made for disposal of sullage water.

4. Dry Toilet Options

All dry toilets deposit excreta into some form of receptacle located below the toilets. The options for receptacle are as follows:

(i)   **Pits.** Pits retain fecal solids while allowing urine and any water added to the pits to soak into the ground. The photos on page 89 show typical pit latrines used in schools. Pits may be closed and replaced by a new latrine when full. Pits located beneath permanent structures, on the other hand, require occasional desludging, with removed sludge transported to place where it can be treated or safely disposed of. This can be problematic, particularly when the sludge has a high solids content, which is usually the case for dry systems.

(ii)   **Vaults.** Vaults perform a similar function to pits but are located partly or wholly aboveground. Their largely aboveground location means that vaults are often easier to desludge than pits, but toilets with vaults will be difficult to heat because of the flow of cold air through the vaults and the toilet holes into the building interior.

---

5   A 60 LPCD is lower than the figure sometimes suggested. In practice, the minimum water consumption figure compatible with the use of sewers depends on the system layout, sewer slopes, and the quality of the sewer construction.

(iii) **Containers.** Containers are removed at intervals of a few days to a week and replaced by a new container. Removed containers must be transported to a suitable disposal or treatment facility, where they are emptied and then cleaned for reuse. Because humans produce more urine than feces, container-based sanitation systems work best in conjunction with UDDT toilets. To reduce odors, users are encouraged to add earth or sawdust to the container contents after defecating.

Dry systems do not deal with gray or sullage water, which has to be disposed of separately.

Identifying an Appropriate Sanitation Option

Figure A4.4 sets out a logical approach to choosing an appropriate option for a school. The focus is on the whole system, including both the toilet and the arrangements for subsequent management of excreta and wastewater. Figure A4.4 illustrates the following points:

(i)   If the school has an adequate water supply and is close to a public sewer, water-flushed toilets connected to the sewer will be the preferred option.

(ii)   If the school has an adequate water supply but no public sewer is available, the choice is between connecting all sanitation blocks to a treatment plant serving the whole school, or providing a treatment facility or septic tank at each sanitation block. Treatment plant options will only be viable if the power supply is reliable and affordable and appropriate operational and management knowledge and skills are available.

(iii)   If water is available but its quantity and/or the system reliability is insufficient to guarantee successful sewer operation, the best option will be to provide pour-flush toilets connected to septic tanks or on-site treatment facilities. Again, the challenge will be to ensure that a management system appropriate to the selected treatment or disposal option is in place.

(iv)   If there is no piped water system, it will be difficult to ensure long-term operation of water-flushed toilets. Thus, the focus should be on identifying a dry sanitation system that meets user requirements and is manageable with locally available resources. The aim should be to provide high-quality waterless toilets in sanitation blocks, located sufficiently far from school buildings to meet the national minimum requirements. One major challenge will be to make arrangements for safe and effective removal of the sludge that accumulates under the sanitation blocks.

(v)   If funds for new sanitation blocks are not available, the short-term focus should be on improving existing facilities.

Once the overall approach to sanitation provision has been decided, attention can turn to the detailed design of the preferred system.

Figure A4.4: Identifying an Appropriate Sanitation Option

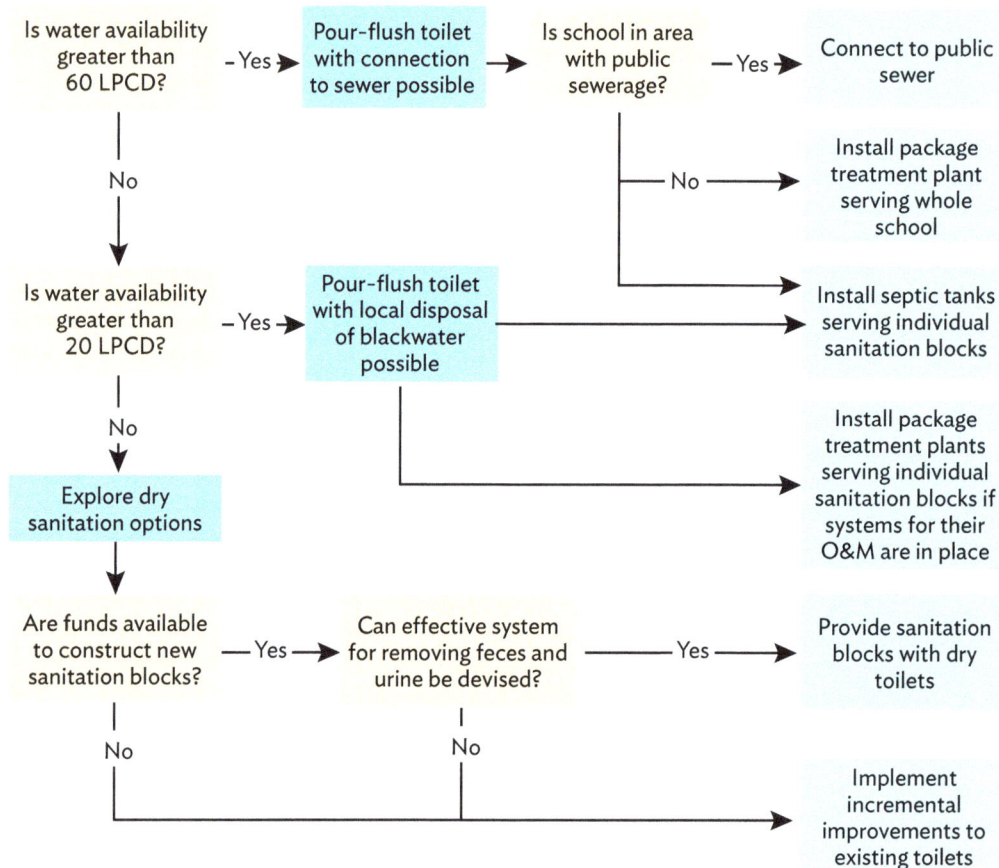

LPCD = liter per capita per day, O&M = operation and maintenance.
Source: Authors.

Immediate Low-Cost Improvements

For schools, especially in rural areas, which have no piped water supply and no prospect of receiving such a supply in the near future, the short-term focus should be on providing improved dry systems located outside of buildings. Since outdoor dry systems do not meet the national minimum requirements, they should be upgraded as soon as resources become available.

Possible measures to improve outdoor toilet facilities include:

(i)   construction of a brickwork "collar" around the upper part of pits to reduce the possibility of pit collapse;

(ii)   replacement of crude board latrine floors with specially fabricated slabs or floor units, including footrests; and

(iii)   replacement of poor-quality superstructures, including replacement of any unlockable doors.

Regardless of which combination of these measures is implemented, the defecation hole should be closed when not in use. The conventional approach is to provide a cover, to be removed by each user before use and replaced after use. The problem with this approach is that users do not always replace the cover, which may be lost or stolen, with the result that the pit is never sealed.

A better option might be to replace open defecation holes with "safe toilet" (SATO) pans. The SATO pan is a simple plastic pan incorporating a counterweighted trap that keeps the pan closed, separating the toilet cubicle from the pit when not in use. When closed, the trap prevents insects from entering the pit and odors from escaping from the pit. When someone uses the toilet, the combined weight of feces and urine overcomes the closing force created by the counterweight and causes the trap to open, allowing excreta to fall into the pit. Once the excreta have cleared the trap, the counterweight causes it to close again. The SATO pan is built into the slab over the pit, which is usually constructed from concrete.[6]

Washroom and Sanitation Block Design

Washrooms and sanitation blocks should provide toilets, handwashing facilities and, where appropriate, shower facilities. They should be well heated, providing a minimum temperature of 15°C at all times, and should be well-lit (both in each toilet cubicle and in public areas) and easy to clean.[7] In addition, they should cater to students with disabilities and special needs.[8] There should be adequate space for toilets and washing facilities, circulation, and storage of users' belongings while they use the facilities. If possible, urinals, toilet cubicles, and handwashing stations should be out of the line of sight for people entering the toilet block (it should be noted that the layouts of boys' toilets as shown in Figures A4.2 and A4.3 do not meet this requirement since the urinals are located immediately opposite the door). Some compromise on such requirements may be necessary when available space is limited.

Regardless of whether the toilet is water-flushed or dry, local preferences should guide the choice between squat and seating-type toilets. The toilet intended for students with disabilities should be of the seat type. Students should be able to open toilet cubicle doors from the inside, while provision should be made for adults to open each toilet cubicle door from the outside in the event that a student inside needs assistance.

1. Layout and Dimensions

The minimum internal dimensions for toilet cubicles should be as follows:

(i)   Standard cubicles should be 850 millimeters (mm) wide by 1,500 mm deep, with inward opening door providing at least a 550-mm clear opening when fully open.

(ii)   Cubicles for students with disabilities require wheelchair access and should be at least 1,525 mm wide by 1,450 mm deep, with handrails on both sides of the toilet and an outward opening door. The corridor outside the cubicle should be at least 1,050 mm wide.[9] For reasons of privacy and security, it should be possible to secure toilet cubicles from the inside.

---

[6]   The SATO system has been used successfully in countries with temperate and hot climates. It should be piloted, and its performance should be monitored before use in countries with extreme cold winters to ascertain whether long periods of subzero temperature would affect its performance.

[7]   North American standards for lighting in restrooms suggest a minimum lighting level of 150 lux.

[8]   The Restroom Association (Singapore) (2018) has more information on the design and maintenance of public toilets.

[9]   Further information can be found in Bobrick (2017).

Toilets should be arranged in rows so that all can connect to a single waste pipe as shown in Figure A4.5. If possible, it is better to arrange two sets of toilets back-to-back so that they can use a common waste pipe. The pipe should be 100 mm diameter and there should be access at both ends so that it can be rodded or flushed in the event of a blockage. While Figure A4.5 shows inspection chambers at each end of the pipe run, it is also possible to replace the chamber at the head of the run with a rodding eye.

Figure A4.5: Typical Toilet Layout Showing Pipework

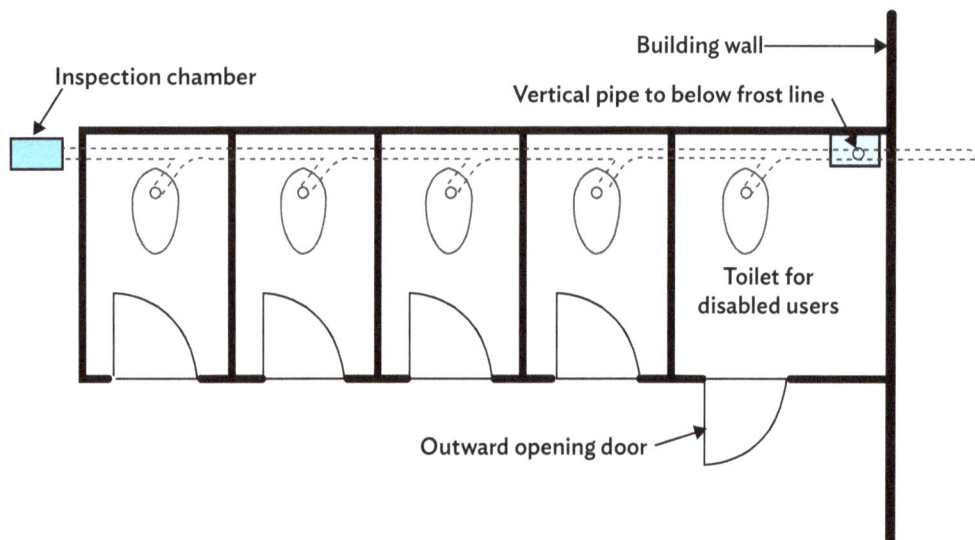

Source: Authors.

Whenever possible, washbasins should discharge wastewater to a common drain pipe. Flows from washbasins and water-flushed toilets should discharge through separate drainage systems within and under buildings, combining at inlet manholes to septic tanks, treatment plants or external sewerage as appropriate. Where flows from pour-flush toilets discharge to a local septic tank, it is also possible to discharge these gray water or sullage flows to a separate soakaway system, which must be located below the depth at which the ground freezes.

2. Heights of Toilets and Washbasins

Washbasins should be available at a height that is accessible to the youngest students. To cater to younger and older students, it may be necessary to provide washbasins at different heights. The following are guideline figures for height:

Children in kindergarten	580 mm
Children in grades 1–3	660 mm
Grades 4–6	735 mm
Grades 7–9	840 mm
Grades 10–12	890 mm

In practice, it is often advisable to compromise and provide a more limited number of washbasin heights, designed to meet the needs of a wider age range of students.

Where toilets are of the seat type, standard toilet seat sizes should be used, with a seat height in the range of 400–430 mm.[10] The seat height for kindergarten children should be in the range of 250–300 mm.

3. Water Supply to Sanitation Blocks

The preferred water supply option for sanitation blocks is a piped supply to an elevated storage tank, which supplies water to handwashing stations, washbasins, and shower facilities. When designing piped water systems, the possibility of freezing must always be considered. Storage tanks should be located within heated buildings and should be well insulated. Pipes from storage tanks to water outlets should run within heated buildings. External pipework should be laid below the depth to which frost penetrates. Particular care is required for the pipes that connect deep external pipework to the pipework within the building to avoid freezing. Appendix 3 provides further information on protection against freezing.

Where there is no piped supply but a water source is available reasonably close to the school, the possibility of transporting water in a tank mounted on a truck, tractor-hauled trailer, animal-drawn cart, or handcart should be investigated.

WHO (1997) suggests the following options for handwashing facilities that do not have a piped water supply:

(i)   a pitcher of water and a basin (one person can pour the water for another to wash their hands; the wastewater falls into the basin);

(ii)   a small tank, for instance a 200-liter oil drum, fitted with a tap, set on a stand, and filled using a bucket, with a small soakaway or a basin under the tap to catch the wastewater; and

(iii) a "tippy-tap" made from a hollow gourd or plastic bottle that is hung on a rope and that pours a small stream of water when it is tipped.

The first and third options use small water vessels, which require regular refilling. This suggests that neither works without additional water storage which might be provided by a small tank or one or more 200-liter drums, as described in the second option. Given the limited capacity of water transport systems, the challenge with all of the options is to ensure that water is used sparingly, with minimal wastage. In schools without a reliable piped water supply, hygiene promotion campaigns should include messages on the need to use scarce water carefully.

For pour-flush toilets, a suitable water container should be placed in each toilet compartment, along with a smaller bowl or jug, to be used to scoop water out of the container for flushing the toilet. The size of the main container might be 40–100 liters while the scoop might have a capacity of around 3 liters.

4. Gender-Related Factors

Toilet facilities for girls should be completely separate from those for boys and have different entrances. The number of toilets for girls should be equal to, or preferably exceed the total number of toilets and urinals provided for boys.[11]

Regular cleaning of toilets is important in all circumstances, particularly for school toilets used by girls. Toilets for post-puberty girls should provide a covered container within each toilet compartment to allow for the disposal of

---

[10]   Further information on dimensions for sanitary facilities for schools is in DODEA (2012). Information on handrail heights and other details of toilets for students with disabilities can be found in Department of Education and Skills (2014).

[11]   The Restroom Association (Singapore) (http://www.toilet.org.sg/articles/GuideBetterPublicToilet.pdf) suggests five toilets for girls to every one toilet and two urinals for boys.

sanitary pads. In addition, regular cleaning of toilets to remove any traces of menstrual blood is essential, not least for dry toilets, where the lack of water makes it harder to maintain hygienic conditions. A brush and a little water should be provided in each toilet compartment for use in cleaning the bowl or the area around the defecation hole.

The Women in Europe for a Common Future (2006) further recommends that every toilet room used by women should be supplied with the following:

(i)   a poster describing the proper cleaning methods for the toilets;
(ii)   a bucket of water with a scoop and a brush for cleaning traces of feces or menstrual blood;
(iii)   paper bags or wrapping material for sanitary napkins and/or tampons; and
(iv)   a bin for disposing of sanitary napkins and/or tampons.

Moreover, the caretaker of the toilet room should be supplied with

(i)   vinegar or another acidic biodegradable agent for the urine bowl;
(ii)   soda water or other biodegradable detergent for the other parts of the toilet; and
(iii)   toilet brushes.

These requirements, modified as necessary to suit Mongolian conditions, provide a guide for toilets used by girls who have reached the age of puberty.

5. Construction Options for Washrooms and Sanitation Blocks

One option for building sanitation blocks fairly cheaply is the container type internal WASH facility piloted in Ulaanbaatar.[12] The toilet block is housed in an adapted shipping container. The concept of the "container house" with toilet and washing facilities originated with mining companies working in remote areas of Mongolia.[13] The key requirement in Mongolia is that the container is insulated, which requires expenditure on insulating material. Information on insulation materials and techniques is available from a number of sources.[14]

Where a sanitation block is connected to a school building through a connecting corridor, it should be well insulated and preferably be provided with some form of heating. To prevent any smells travelling through the corridor to the main building, there should be two doors between the toilet block and the main building, one at each end of the connecting corridor. The doors should be kept shut to keep smells out of the main buildings and heat in both the main building and the sanitation block.

Sewerage and Drainage Design

Pipework runs within insulated buildings do not require either heating or additional insulation, provided that the heating system can be relied on 24-7 throughout the winter. If it is nonoperational for even short periods in winter, pipes located within buildings freeze. This is likely to be of particular concern for pipes that run through sanitation blocks that are isolated from the main school buildings.

---

[12]   Kindergarten no.23 in Nalaikh District. It should be noted that the term "container" has a different meaning here than when used in the context of "container-based sanitation."

[13]   The concept is also used in other countries. For instance, Structure Now Social Solutions (http://www.structurenow.com/ablution-toilets.php).

[14]   For example, see the Building Science Corporation's information sheet BSI-031 Building in Extreme Cold (Lstiburek 2010), and resources of the Cold Climate Housing Research Centre (http://cchrc.org/).

As for water pipes, drainage pipes that run outside heated buildings must either run below the deepest depth of frost penetration or be provided with permanent heating. Although the capital cost of laying pipes below the frost penetration depth is high, it is usually a better option than relying on pipe heating systems which must operate 24-7 during the winter months, with high operational costs. Without heating for extended periods, pipes freeze, resulting in breaks in system operation and broken pipes which leak untreated wastewater when the system thaws. Experience in many parts of Mongolia, in schools and health facilities, shows that such system failures caused by freezing lead to the abandonment of "improved" facilities, with people reverting to use of old unsatisfactory external pit latrines.

Drainage pipes connecting toilets to a septic tank or sewer should normally be 100 mm diameter. Connection pipes from washbasins can be smaller, either 38 mm or 50 mm diameter, depending on the number of washbasins connected to the pipe. Sewers that connect toilet blocks, washbasins, and laundry facilities to local treatment facilities should not be less than 100 mm diameter.

The minimum slope of the pipe collecting flow from a single row of toilets should normally not be less than 1 in 40 (2.5%). Sewers collecting flows from whole buildings or toilet blocks may be laid at shallower gradients, if possible, at a slope of 1 in 100 (1%) but not less than 1 in 150 (0.667%).

Manholes should be provided at junctions and changes in direction and/or slope. It is advisable to use a plastic film wrapped around the manhole to prevent bonding of the soil to the structure, and thereby preventing damage from frost heave. It is also advisable to insulate each manhole with a minimum of 75 mm of polystyrene or polyurethane encased within plastic. To further reduce cold penetration into the manhole, an insulated removable cover should be provided in the manhole itself.

Wastewater Treatment and Disposal Options

A sewer connection is the best option for schools that are located close to a public sewer. In all other cases, some form of on-site treatment and disposal, with occasional removal of sludge, will be required. The options are to provide either a single treatment facility to serve the whole school or separate treatment facilities for individual sanitation blocks or washrooms.

1. Treatment and Disposal Options

Treatment options fall under the following broad categories:

(i)   Anaerobic Treatment Facilities and/or Processes
  - Septic tank providing 30%–40% organic load reduction and up to 90% (1 log) pathogen reduction,
  - Baffled reactor providing 50%–60% organic load reduction and similar pathogen reduction to septic tank, and
  - Upward-flow anaerobic filter often included in the last compartment of a baffled reactor with similar organic load and pathogen reduction

(ii)   "Extensive" or "Natural" Treatment Processes
  - Waste stabilization ponds, including anaerobic ponds, facultative (partly anaerobic), and maturation (fully aerobic) ponds that can achieve good results, but will need a lot of space and are likely to freeze in winter.[15]

---

[15]   Some *aimag* centers have waste stabilization pond treatment, which does not appear to freeze. This might be partly explained by how sewage has been heated where sewers run alongside heating pipes.

(iii)   Conventional Aerobic Treatment Processes
  - Mechanized package treatment plants including plants based on activated sludge process, such as sequencing batch reactors; moving bed biological reactor (MBBR) treatment processes; and rotating biological contactors.

Septic tanks are used to serve individual sanitation facilities, with effluent discharged to a soakaway or drainfield and relying on percolation through the ground to reduce pathogen concentrations to safe levels. Baffled reactors and upward-flow filters may be used to serve individual sanitation facilities or the school as a whole. In theory, there is no reason why they should not discharge effluent to a soakaway or drainfield, although they provide no obvious advantages over septic tanks when deployed in this way. Waste stabilization ponds should only be considered as an option for treatment of wastewater from the school as a whole.

Package plant options should normally be deployed at the level of the school as a whole. If operated effectively, all plants will achieve a higher reduction in organic load than anaerobic options although none will reduce pathogen concentrations to levels that are safe for discharge to a watercourse without some form of disinfection. Rotating biological contactors should be preceded by a tank or septic tank. Activated sludge, sequencing batch reactor and MBBR processes require sedimentation after the aeration stage in a separate tank, included as part of the package plant. Sequencing batch reactors use the same processes as activated sludge plants but operate in batch mode with aeration and sludge settlement taking place in the same reactor rather than in separate units. All package plant options rely on a reliable power supply, and require knowledgeable operators and skilled maintenance technicians. There is a high probability that these plants will not be operated consistently because schools cannot afford their high running costs. Box A4.2 provides a short case study of a MBBR treatment plant installed at a kindergarten in Darkhan-Uul Aimag.

The rotating biological contactor system requires less energy than other mechanical treatment options. However, the bearings of the rotating contactors are susceptible to failure and will be difficult to repair when located in an underground chamber. Other aerobic options use bubble diffusion with air supplied by a compressor. The compressor can be located in a building with air piped to the treatment unit, which facilitates maintenance.

Box A4.2: Moving Bed Biological Reactor Treatment Plant at a Kindergarten

A moving bed biological reactor (MBBR) treatment plant was installed to serve a kindergarten in Orkhon Soum in Darkhan-Uul Aimag. The contract for this plant required the contractor to operate and maintain the plant for the first 2 years after commissioning. After the 2-year maintenance period ended, the responsibility for operating and maintaining the plant lay with the kindergarten management. Key operation and maintenance (O&M) tasks included maintenance and repair of mechanical equipment and desludging.

Although the MBBR technology is claimed to generate less sludge than a conventional activated sludge system, there is still a need for desludging at intervals of roughly 2 years. This was difficult for the kindergarten management. A possible solution could be to install similar mechanized treatment facilities in several schools in the same area and contract a private sector organization to operate and maintain those facilities. The initial contract period could be 3 years with the possibility of longer contract periods if the initiative is a success. The feasibility of this option depends on the

  - availability of contractors with the knowledge and experience required to carry out the required O&M tasks at an affordable price; and
  - ability of government authorities to formulate, effect, and monitor contracts.

continued on next page

Box A4.2 *continued*

This option is most likely to be viable in peri-urban areas. Because of the distances involved, it is unlikely to be feasible in remote rural settlements. The key point is that mechanized systems should not be implemented unless a clear and workable plan for their long-term O&M has been formulated.

Source: J. Ilias and B. Scharaw. MOMO Fact Sheet: Decentralized Waste Water Management: Experiences from Pilot Operation in Orkhon Sum, MOMO Integrated Water Resources Management. http://iwrm-momo.de/download/MoMo%20Fact%20Sheet%20WWTP%20Orkhon%20Sum.pdf.

According to building regulations in the United Kingdom, any treatment facility must be located at least 7 m from the nearest habitable property and that soakaways must be located at least 10 m from a water source, 15 m from a building, and 50 m from a watercourse. Where treated effluent is discharged to a watercourse, the discharge point must be at least 10 m from the nearest habitable building. If discharge is to a soakaway, a sampling chamber should be provided prior to the soakaway.[16]

2. Protection against Freezing

All the wastewater treatment options mentioned require protection against freezing. For septic tanks, baffled reactors, anaerobic filters, and package plants, the best way to provide protection is to bury treatment facilities at a sufficient depth to avoid freezing. In addition, heat loss can be reduced by providing an insulating layer at least 100 mm thick on the top and down the sides of septic tanks and buried package treatment plants. Polyurethane or polystyrene might be used as the insulating layer. Polystyrene, however, must be of a type suitable for installation beneath ground level.

The burial option is not possible for facultative and maturation waste stabilization ponds, which rely on sunlight for their operation. North American design guidelines for waste stabilization ponds in cold climates recognize that ponds freeze in winter, with minimal biological activity. It is common practice in Canada and the northeastern United States to design ponds to retain all wastewater generated through the winter months and then to allow controlled discharge in the late spring and early autumn. North American ponds are designed to provide at least 1.5 m depth below the ice layer. Waste stabilization ponds may not be possible in Mongolia and are unlikely to be the best option for school wastewater treatment because of the long winter season, large land take, and potential operational difficulties.

---

[16]   Cited in WTE Ltd. (https://www.wte-ltd.co.uk/wastewater_legislation.html).

3. Septic Tanks and Baffled Reactors

Septic tanks normally have two compartments, the first about twice as large as the second. Figure A4.6 shows a typical septic tank arrangement suitable for use in cold climates. The whole septic tank is buried at sufficient depth to avoid freezing. Deep manholes allow access.

Figure A4.6: Typical Septic Tank to Drainfield Arrangement

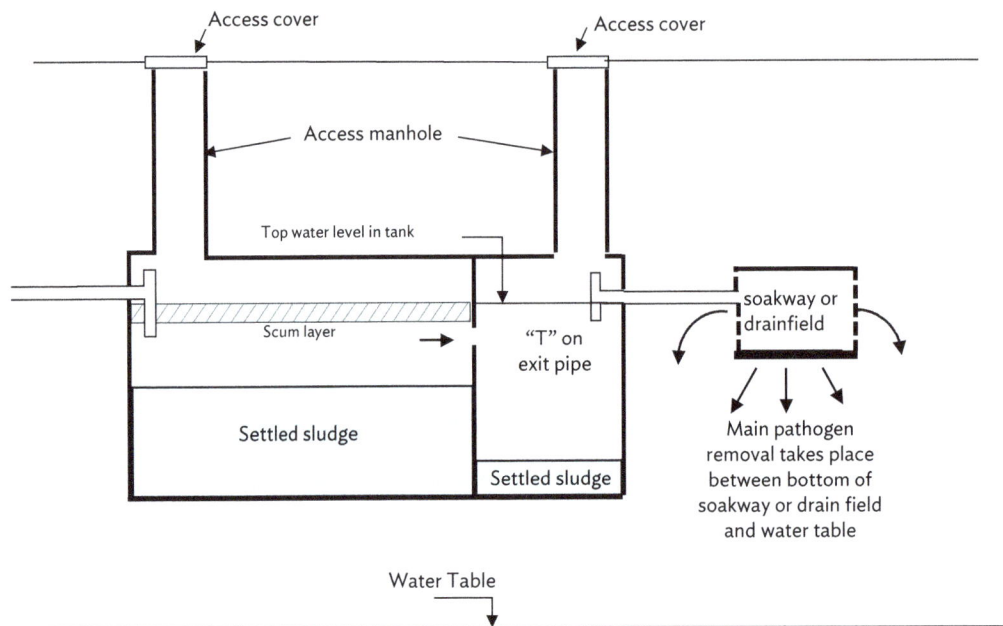

Source: Authors.

Septic tanks work through sedimentation of settleable solids and subsequent digestion of those solids. As solids digest, some material rises to the surface, where it remains in the form of scum. Relatively clear water passes through the septic tank and discharges either to a soakaway or drainfield system or to secondary treatment. The standard two-compartment design is intended to reduce the possibility of short-circuiting, although some researchers have found no difference in performance between single and double compartment tanks.

Septic tanks must provide sufficient capacity to accommodate sedimentation of settleable solids, and digestion of settled solids and subsequent storage of digested solids and scum. Table A4.1 gives calculated volumes required for sedimentation and for sludge accumulation for a range of populations, assuming typical figures of 50 LPCD wastewater flow, a sludge accumulation rate of 60 liters per capita per year (l/c.yr) and desludging at intervals of 5 years.[17] Given the low ambient temperatures, a higher allowance for sludge accumulation is required in cold climates such as in Mongolia.

---

[17]   The estimates for retention time and sedimentation volume start from the figures given in the guidelines produced by the Associação Brasileira de Normas Técnicas (ABNT), the Brazilian organization that sets technical standards (ABNT 1982).

Table A4.1: Calculations for Required Septic Tank Volume

Population	Sedimentation		Sludge accumulation		Total volume required (m³)
	Retention time (h)	Retention vol (m³)	Rate (l/c.yr)	Volume (m³)	
15	24	0.75	60	4.5	5.25
30	24	1.50	60	9.0	10.50
45	24	2.25	60	13.5	15.75
60	24	3.00	60	18.0	21.00
75	24	3.75	60	22.5	26.25
100	24	5.00	60	30.0	35.00
150	24	5.94	60	45.0	50.94

h = hour, l/c.yr = liter per capita per year, m³ = cubic meter, vol = volume.
Source: Associação Brasileira de Normas Técnicas (1982).

The overall length-to-width ratio of septic tanks is typically around 3:1. Using the total volumes given in Table A4.1 and assuming a liquid depth of 2 m, a septic tank to serve 75 people should have plan dimensions of about 2.1 m breadth by 6.3 m width. A prefabricated tank might be considered in place of a rectangular tank, as prefabrication reduces the time required for septic tank installation.

Septic tanks are a form of primary treatment and do not normally reduce the organic load of wastewater sufficiently to meet standards for effluents discharged to surface water bodies. Nonetheless, the organic load is less important when discharge is to a soakaway or drainfield, as most septic tanks are designed as part of a septic tank–drainfield system. Figure A4.7 is a diagrammatic view of such a system with the ground cut away to show the septic tank and drainfield. It should be noted that both the septic tank and the drainfield have to be deeper in Mongolia.

Figure A4.7: Septic Tank to Drainfield

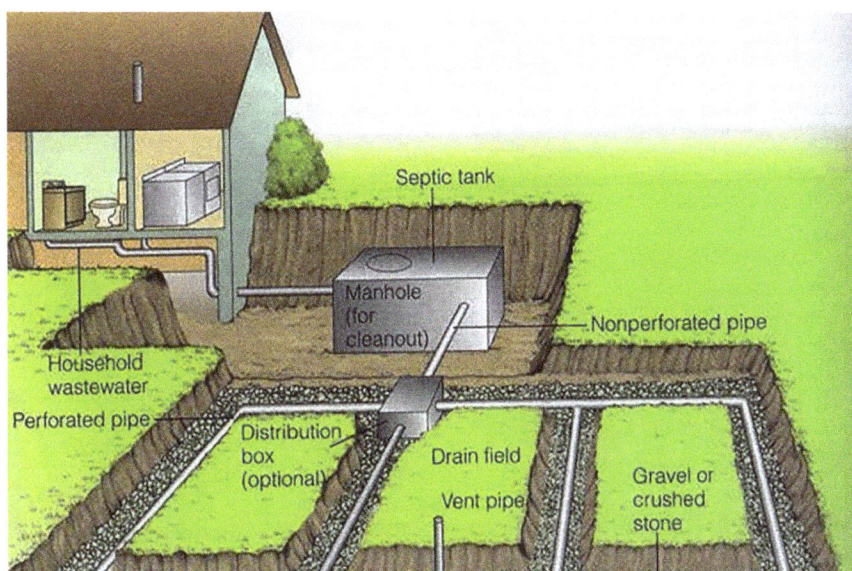

Source: Instructables. https://www.instructables.com/id/How-to-Choose-a-Septic-Tank-Portland-Specialist/.

The drainfield should be located on a site with a slope of less than 25%, preferably less than 10%. When the slope is greater than 10%, the drainfield trenches should be laid across the slope. The bottom level of the drains should be at least 1.5 m above the highest level reached by the water table. The size of drainfield required will depend on the flow and the permeability of the soil.

A well maintained anaerobic baffled reactor can remove up to 20% more organic load than a simple septic tank, providing what might be termed "enhanced primary treatment." Further treatment is normally required before discharge to a lake, river, or stream. Figure A4.8 shows the configuration of a typical anaerobic baffled reactor, comprising a "settler" compartment, similar to the first compartment of a septic tank, followed by a series of shorter compartments. These are configured to force the flow in at the bottom of the compartment and out at the top of the compartment. The theory is that solids are removed from the flow as it passes through the sludge at the bottom of each compartment, with the sludge fulfilling the same function as the sludge blanket of an upward-flow anaerobic sludge blanket reactor.

Figure A4.8: Typical Anaerobic Baffled Reactor

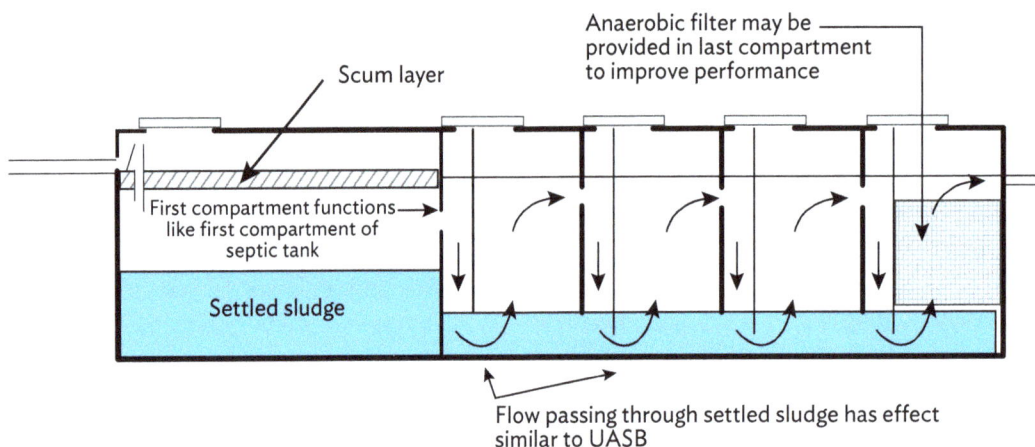

UASB = upward-flow anaerobic sludge blanket reactor.
Source: Authors.

The challenge with baffled reactors, as with simple septic tanks and other decentralized treatment systems, is to remove sludge before it builds up to a point at which it compromises the operation of the reactor.

4. Package Plants

Many manufacturers produce prefabricated package plants that are assembled in the factory and delivered to the site for installation. Most prefabricated package plants involve some variant on the activated sludge process. They require a reliable power source, a trained operator, and back-up services for desludging and dealing with mechanical failures and other problems that are not solvable at the local level.

5. Gray Water

The arrangements for gray or sullage water disposal depend to some extent on the type of sanitation adopted. Where toilets are water-flushed, gray water can be discharged to a manhole or chamber that also receives black water flows from toilets via a separate drainage system. Another option, which may be suitable where water is scarce, is to use gray water to flush toilets, by providing a storage tank for gray water, from which water can be taken in a container to flush pour-flush toilets. The gray water supply should not be linked directly into toilet cisterns as back-siphonage might cause it to mix with clean drinking water.[18]

Gray water should not normally be stored for more than one day before use or disposal. The storage time can be increased if chlorine bleach is added to the gray water at a rate of around one teaspoon for every 2 liters.

Dry Sanitation Options

In compliance with the national minimum requirements for WASH in schools, dry toilets must be located at least 20 m from other school and dormitory buildings, well insulated, and provide adequate ventilation and lighting. The sanitation blocks containing dry toilets should provide handwashing facilities for users to wash their hands after defecation, even if water has to be carried from the water source.

There are general lessons to be learned from pilot sanitation blocks that contain dry toilets installed in schools in Mongolia. For example, the evaluation report on a WASH in schools and kindergartens project (Alcanz Consulting Group 2015) cites two problems with the sanitation blocks containing dry toilets piloted under the project (photo):

(i)   the high cost of operating the electric heaters used to warm the sanitation block interiors, and

(ii)   the difficulties experienced in emptying pits once they were full.

Sanitation blocks with dry toilets. *The blocks are well insulated and ventilated with lighting* (photo by Alcanz Consulting Group 2015).

Electric heaters are usually more expensive to operate than other forms of heating. The report notes that the high cost of running the heaters has led to some schools shutting off their heaters at night, resulting in condensation problems and damage to paneling and the heaters themselves. The other major disadvantage of electric heaters is their dependence on a reliable 24-7 electricity supply, which may be difficult to guarantee in remote *soums*.

A better approach to heating is to site the sanitation blocks as close to the school and dormitory buildings that they serve as is allowed by the minimum requirements for WASH in schools, and to run heating ducts from the

---

[18]   For further information, see Greywater Action (http://greywateraction.org/systems-for-cold-climates-including-wetlands/).

buildings to the sanitation blocks. Although this arrangement may result in some loss of heat along the heating duct, its costs should be more than balanced by the fact that all heat generation can be concentrated at a small number of locations. Provision of a connecting corridor between a main school or dormitory building and the sanitation block removes the need to go out into the cold. The heating pipe or duct from the main building to the sanitation block can then be run along or underneath the corridor. Figure A4.9 shows this arrangement diagrammatically with two sets of doors at the end of the main building and the corridor to prevent smells and flies passing along the corridor and into the main building.

Figure A4.9: Sanitation Block with Connecting Corridor

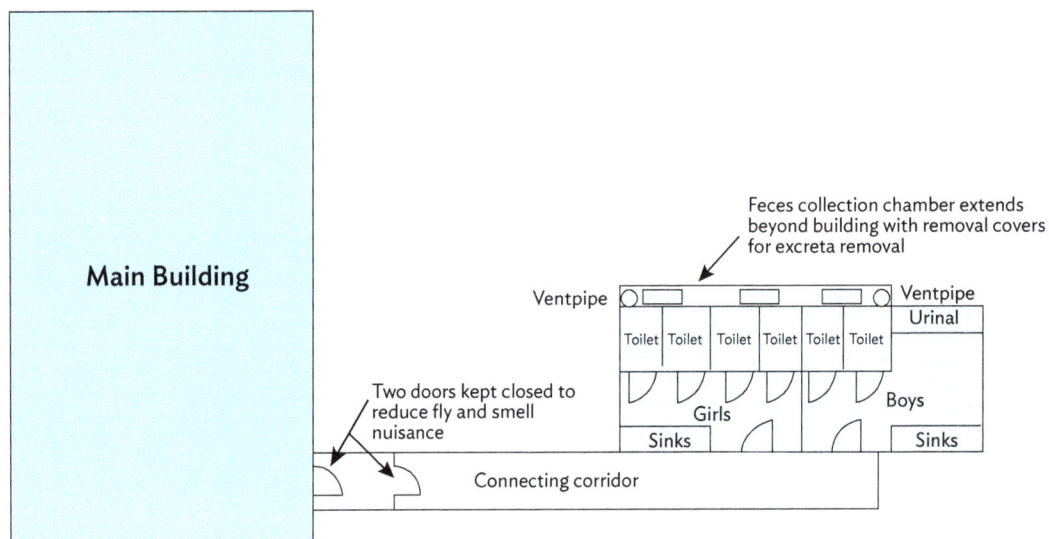

Source: Authors.

The dry toilets piloted in Khuvsgul Aimag are nominally of the ventilated improved pit (VIP) latrine type, incorporating vent pipes topped with wire mesh located outside the sanitation block structure. The theory of the VIP latrine is that a combination of heating of the external pipe and the flow of air across the top of the pipe encourages a flow of air from the interior of the sanitation block, through the pits and up through the vents to the atmosphere. This vents odors out of the sanitation block interior and forces flies and other insects to fly up the vent pipe rather than back into the sanitation block interior. Unfortunately, the mechanism breaks down in cold climates as heavier cold air reverses the flow through the vent pipe.

To reduce odors, flies, and other insects, a better approach may be to ensure that the defecation holes are closed except when a toilet compartment is used. This can be achieved by installing SATO pans, as described on page 99. Use of SATO-type toilets might allow sanitation blocks to be located closer to school and dormitory buildings than the 20 m specified in the national minimum requirements. This would both enhance the convenience of users and reduce costs.

The challenge of desludging must be overcome if dry toilets installed in a permanent sanitation block are to continue to function over time. Because dry sanitation relies on direct-drop toilets, the pits must be located beneath the structure. The normal arrangement for such latrines is to extend the pit beyond the back wall of the structure and provide a removable slab for access, as shown in Figure A4.9. The solids content of the sludge that accumulates in dry pits is high, making it difficult to remove. Where a vacuum tanker is available, adding water to the pit contents may change their characteristics sufficiently to allow them to be pumped from the pit. Where a vacuum tanker is not available, adding water is difficult due to lack of water, or the sludge fails to liquify sufficiently to be pumped, manual emptying will be the only removal option. In *ger* areas of Ulaanbaatar, workers often remove sludge from pit latrines during the winter when the sludge is frozen, which makes the task less unpleasant for the workers.

Raising sanitation blocks above ground level and replacing pits with vaults reduce the difficulties associated with sludge removal. One major disadvantage of this system is that it allows cold air to enter the superstructure through the squatting holes. The use of self-closing SATO-type toilets would reduce but not eliminate the effect of cold air entering the superstructure from below.

The difficulties associated with sludge removal could be further reduced by implementing a well-designed and managed container-based sanitation system. One container-based sanitation initiative in Ulaanbaatar operated successfully for a time with support from an international nongovernment organization.[19] With the end of the funding support, however, the initiative ceased.

The main challenges presented by the container-based sanitation approach are to

(i)   develop an effective container design to ensure that the container is easily transportable, easy to empty, and does not spill its contents during transport;

(ii)   provide an acceptable disposal and treatment option; and

(iii) cover costs, particularly the cost of regular replacement, removal and emptying of containers, and subsequent cleaning of the containers and treatment or safe disposal of the container contents.

Several container-based sanitation designs exist, including the Mosan toilet piloted in Bangladesh, the Sanergy Fresh Life toilet in Nairobi, Kenya, and the Ghanasan toilet in Kumasi, Ghana, as described in Box A4.3

Box A4.3: Container-Based Sanitation Designs

The Mosan toilet is a urine-diverting dry toilet that was developed for use in Bangladesh. It is a prefabricated lightweight toilet intended for home use. To facilitate transport, Mosan toilets can be stacked. Feces and urine are separated into 10-liter removable containers, which allow easy and convenient collection and transport. The designers claim that its elegant design has the potential to become a status symbol, raising awareness and creating demand for hygienic sanitation. In view of the small size of the containers, it is likely that the system would have to be modified for multiple user use in a school setting.

The Sanergy Fresh Life toilet is a complete toilet system fitted into a "super container" that uses prefabricated "cartridges" to collect urine and feces. The cartridges are firmly locked into place for transport so that they can be swapped out as they become full and pose no risk to the environment. The Sanergy toilets are franchised to local operators to provide services in low-income areas in Nairobi, Kenya, with each unit typically used by 50–60 people, similar to the optimum number of users for a typical school sanitation unit.

The Ghanasan toilet, developed as part of a pilot project to introduce a franchised sanitation service in Kumasi, Ghana, is an integrated unit, with an upper hinged pan section that can be lifted to allow access to the container. A sealable waste cartridge inserted into the container collects feces. Urine is directed to a pipe and out of the toilet. Operators regularly remove full waste cartridges and replace them with clean cartridges.

Sources: Mosan. http://www.mosanitation.com/; Sanergy. http://www.sanergy.com/our-model/; E3Design Ltd. http://e3design.co.uk/ghanasan--clean-team-toilet.html.

---

[19]   Recent trials of container-based sanitation systems in Nairobi, Kenya also showed promising results (Bohnert et al. 2016).

(further information on container-based sanitation designs can be found in EOSS and Water, Engineering and Development Centre 2014). Almost all incorporate separation of urine and fecal solids. Designs fall under the following broad categories:

(i)   "integrated" designs, in which the toilet pan fits directly onto the container that can be removed and changed by lifting the toilet pan (photo); and

(ii)   designs, in which the container is located in a separate compartment beneath the toilet.

One advantage of the integrated option is that the toilet and container are contained entirely within the sanitation block building so that there is no heat loss to a cold external compartment. Another advantage is that it would be easy to modify existing buildings to incorporate the new toilet designs. The big disadvantage is the relatively small size of the container.

Source: Engineering for Change. https://www.engineeringforchange.org/solutions/product/ekolakay-toilet/.

Existing designs typically use containers with capacities of around 20 liters. Depending on diet, per capita production of feces is typically in the range of 100–400 grams per day (equivalent to roughly the same volume in milliliters [ml]). Urine production volume is higher, typically lying in the range of 800–1,200 ml per person per day. Assuming urine-diversion toilets, with per capita feces production of 200 ml per day and 50 users per toilet unit, the container serving one toilet would fill at a rate of 10 liters per day. Allowing for some variation in per capita feces production and user numbers, a urine-diversion toilet with a 20-liter feces container requires daily emptying. For the same emptying frequency, the required capacity for a urine container would have to be around 100 liters, allowing for some variation in per capita urine amount and number of users.

The main technical challenges with container-based sanitation systems in cold climates lie in dealing with urine and treating fecal sludge. The cold winter conditions rule out the possibility of urine discharge to an external tank, container, or soakaway. Although discharge to a tank located within the structure and to a soakaway system buried at depth is possible, both will fail if the temperature of the buildings falls below freezing for extended periods. This is likely to occur if either the heating system is unreliable, or the funds required for its operation is unavailable.

The experience of the SOIL toilets in Haiti, a container-based sanitation system similar to the Sanergy Fresh Life toilet, shows that cost recovery is an issue for systems serving individual households.[20] For school sanitation facilities, the challenges are to develop a reliable system for regular container removal, emptying, and cleaning; and to provide the funds required to operate this system. Meeting these challenges requires an individual or a small team charged with the day-to-day operation of the system under the overall supervision of the school principal or management committee. The school and *aimag* education department will have to include provision for salaries, materials, and equipment in the annual budget allocated to the school.

Sludge Removal, Treatment, and Disposal Options

Septic tanks, waste stabilization ponds, and decentralized package plants require periodic removal of sludge. Without such regular removal of sludge, their treatment performance will deteriorate, leading to eventual failure. There is evidence that the lack of vacuum trucks to desludge septic tanks has led some schools to limit use of indoor flush toilets. To overcome this constraint, schools must have access to sludge removal services, and provision must be made for sludge treatment. This is true regardless of the form of decentralized treatment, including those that minimize the amount of sludge produced (it should be noted that no treatment system completely eliminates sludge production).

Vacuum tankers offer an effective option for removing sludge from septic tanks and package treatment plant sedimentation units. However, they are expensive and it may not be justifiable for individual *soum* centers to purchase their own vacuum tankers, as each school only requires their services at infrequent intervals, typically every 2–5 years. The aim should therefore be to ensure that vacuum tankers are available at *aimag* centers or other central locations, for use by schools, other institutions, and individual households as and when required.

Low-technology options for treating fecal sludge include the following:

(i)   dewatering on unplanted drying beds;
(ii)   dewatering on planted drying beds;
(iii)   biodigestion in a domed or geobag type biodigester;
(iv)   co-composting with a suitable organic material; and
(v)   lagoons, which is mainly an option for septage with a relatively high water content rather than fecal sludge taken from dry toilets.

Dewatering on unplanted drying beds is a simple option. Freezing during the winter months and subsequent thawing should result in reduced sludge volume and lower pathogen concentrations. Horizontal flow constructed wetlands have been used successfully in some parts of Canada and the northern United States. However, their relatively high management requirements and uncertainty about their efficacy in extreme cold climates mean that they are unlikely to be a viable option for countries like Mongolia. Similarly, it is doubtful whether biodigestion is a practical option in extreme cold climates.

Successful municipal co-composting exists in Fairbanks, Alaska, where the average winter temperature in January is −22°C. In Ulaanbaatar, a project of the nongovernment organization, Action Against Hunger, successfully demonstrated co-composting of fecal material removed from pits in ger areas. However, co-composting is a

---

[20]   The experience of the SOIL toilets show that their container-based sanitation system recovers only about 30% of its operational costs. More information on the SOIL toilets is in https://www.oursoil.org/.

relatively labor-intensive sludge treatment option. The project ran only as long as external funding was available to support the operating costs of co-composting.

Dry latrines, vaults, and container-based sanitation systems produce a relatively small volume of sludge with a high solids content. Therefore, the best option is to discharge the sludge to drying beds which can take the form of shallow basins, up to 1 m in depth. The plan dimensions of beds depend on the volume of sludge produced, with individual beds typically being around 10 m long by 5 m wide. It is appropriate to discharge more than one load to each bed until the depth of sludge in the bed reaches 200–300 mm, possibly more immediately before the winter.

At the *soum* or *aimag* level, the best sludge treatment or disposal option for liquid sludge is to discharge to shallow lagoons managed by appropriate local government authorities, which is a simple option benefiting from freeze–thaw mechanisms. Lagoons should have the following characteristics:

(i)   at least 1 kilometer from the nearest development;
(ii)   located on a flat site, clear of stormwater drainage paths;
(iii) located well away from any drinking water source; and
(iv)   up to 2 m in depth (the depth of sludge in the lagoon may be less).

Lagoons should typically be up to 30 m long by 10 m wide, and should be arranged in parallel so that one lagoon can be left to dry out while another is in use. Where the inflow of sludge exceeds the combined percolation and evaporation capacity, the quality of the effluent can be improved by providing a second lagoon in series. Lagoons freeze during the winter, which facilitates sludge dewatering.

Because of the relatively small amount of sludge and the likely absence of motorized sludge removal and transport vehicles, the sludge from dry latrines may need to be dealt with close to the school. The disposal site should be fenced off and located downwind of the school and other developments, as far from the school as is practically possible and well away from the school water supply, in an area where the water table is not close to the surface. The simplest disposal option is to dig a pit, deposit fecal sludge into the pit, and cover the sludge with a 100–150 mm deep layer of soil. Alternatively, if the sludge is dry and the site is sufficiently far from school buildings and other inhabited buildings, it is possible to stack the sludge above the ground, mixing it with solid waste. After storage for at least a year, the combined effects of solar radiation during the summer and freezing during the winter should reduce pathogen concentrations in the sludge to safe levels. The sludge can then be used as a soil conditioner, if demand exists.

APPENDIX 5
Hygiene Education Approaches

Over the years, several hygiene education methodologies have been developed, publicized, and used. Appendix 5 introduces some tested methodologies to teachers, social workers, school doctors, school management, and education administration at the local and national levels for planning and conducting hygiene education activities. It includes approaches to identify the risk practices of students and teachers and beliefs behind those practices, and to facilitate the adoption of alternative practices that reduce the risk and that are feasible in the local situation. This appendix supplements the information contained in section VI.

Knowledge, Attitudes, and Practices Survey

Hygiene education activities are more likely to be effective if they build on what students and teachers already know, recognize what they think and do, and respond to what they want. It is therefore essential to understand existing knowledge, beliefs, and practices among students and teachers, which can generally be achieved by asking simple questions as follows:

(i)   What do students and teachers believe and what do they know?

(ii)   What do they do?

(iii)   Why do they do it?

(iv)   What barriers prevent them from adopting "better" practices?

Observation can also lead to an understanding of what students and teachers do and the barriers to adopting better hygiene practices. Observation can focus on the following questions:

(i)   Is water available at or near toilets and dining areas?

(ii)   Is soap also available?

(iii)   Do students and teachers wash their hands after using the toilet and if so, do they use soap?

(iv)   Do students and teachers wash their hands before eating and if so, do they use soap?

When observation reveals differences between what students and teachers say that they believe and know, and what they actually do, informal discussion groups with students and/or teachers can help gain an understanding of why they act as they do, and the barriers they face when adopting better practices. Knowledge, attitudes, and practices (KAP) surveys can provide more detailed answers. When assessing the results of such exercises, it is important to "triangulate" information by comparing the information obtained from different sources and in different ways.

KAP surveys use predefined questions contained in standardized questionnaires to obtain quantitative and qualitative information. They can reveal the attitudes and beliefs that are likely to create barriers to behavioral change. However, it should be noted that KAP surveys can only produce useful and reliable results if they are well-designed, carried out by appropriately trained surveyors, and field-tested.[21]

A KAP survey may cover the following topics:

    (i)   sources of students' knowledge about water, sanitation, and hygiene (WASH),

    (ii)   students' knowledge about WASH,

    (iii)   students' perception about sanitation and waste management,

    (iv)   students' use of WASH facilities at school,

    (v)   students' sanitation and hygiene practices outside of school,

    (vi)   teachers' perceptions about WASH at school,

    (vii)   teacher training on hygiene education, and

    (viii) teacher's views of students' hygiene practices at and outside of school.

It is always advisable to analyze the results of surveys, and use the results to modify future surveys. Box A5.1 lists main lessons drawn from the experience of a nongovernment organization with conducting KAP surveys.

Box A5.1: Lessons from Knowledge, Attitudes, and Practices Surveys

Action Against Hunger is an international humanitarian organization that aims to end hunger, treat and prevent malnutrition, and improve access to clean water. The experience of Action Against Hunger with knowledge, attitudes, and practices (KAP) surveys illustrates how even a few small errors can make a survey useless, and how important it is to adhere to simple best practices.

Key lessons from their experience include the following:

    (i)   It is important to choose an appropriate sampling method, understand its limitations, and implement it properly.

    (ii)   Questionnaire design and testing are key requirements in making a survey fool-proof.

    (iii) Training of the survey teams must be thorough, and the survey leaders must make special efforts during the survey exercise to ensure quality control of data collection from the beginning to avoid unwanted surprises in the end.

    (iv) It is necessary to follow basic best practices for data analysis, presentation, and reporting, and to think critically about calculations and analysis and the conclusions drawn from them.

    (v)   The possibility of errors in surveys cannot be discounted, and when an error is discovered, open discussion of that error and the options for addressing it quickly is essential.

Source: Action Against Hunger International. 2013. Conducting KAP Surveys: A Learning Document Based on KAP Failures. https://www.actionagainsthunger.org/sites/default/files/publications/ACF_Conducting_KAP_Surveys_Jan13.pdf.

---

[21]   An example of KAP surveys on WASH in schools is in UNICEF and REACH (2018).

A full KAP survey, covering all aspects of hygiene promotion, requires considerable human and financial resources. Similarly, follow-up hygiene education activities have considerable resource requirements, which may not be available at the school or *aimag* level. Under such circumstances, the focus should be on promoting a small number of the most important hygiene practices.

Participatory Hygiene and Sanitation Transformation

The best known approach to sanitation and hygiene behavior change is Participatory Hygiene and Sanitation Transformation (PHAST). This approach was developed based on the participatory methodology recognizing people's self-esteem, associative strength, resourcefulness, action planning, and responsibility (SARAR) to contribute effectively to decision-making and planning for meaningful change. Box A5.2 summarizes the PHAST approach.

Box A5.2: PHAST Approach to Sanitation and Hygiene Behavioral Change

Participatory Hygiene and Sanitation Transformation (PHAST) starts from the assumption that participatory activities, designed to increase people's awareness of their water, sanitation, and hygiene situation, could empower people to develop and carry out their own plans to improve their situation. It involves the following seven steps:

(i)   problem identification,

(ii)   problem analysis,

(iii) planning for solutions,

(iv)   selecting options,

(v)   planning for new facilities and behavior change,

(vi)   planning for monitoring and evaluation, and

(vii) participatory evaluation.

The focus is on empowering people to develop and carry out their own plans to improve their situation. The plans adopted may include both construction and management of new physical facilities as well as safer individual and collective behavior.

PHAST uses a series of pictures depicting possible situations. Groups of people are asked to describe how these possible situations relate to the local situation and what they would need to do to solve the problems identified through the discussions generated by various tools. When canvassing opinions, a process called pocket chart voting is used, which allows the participants to vote in secret. The results are then discussed by the group as a whole without individuals having to reveal their choice.

Source: World Health Organization (1998).

The PHAST approach has several limitations in relation to the promotion of improved hygienic practices in schools. The first is its relative complexity: it involves seven steps, each of which includes one to four activities. Various commentators (for instance, Peal, Evans, and van der Voorden 2010) have noted that its success depends on the availability of skilled and experienced facilitators and an intensive management structure. This makes it difficult to implement the approach at scale, especially where resources to support training and facilitation are in short supply, as is the case in Mongolia. Under the circumstances, shortcuts may be taken and the approach could be implemented incorrectly. The PHAST approach is more suitable when the aim is to take a comprehensive approach to behavioral change, while it is less appropriate when the aim is to bring about a small number of key behavioral changes such as handwashing and use of safe water.

Another possible disadvantage of the PHAST approach is its emphasis on empowering people to plan rather than on hygiene promotion. Its relevance to school sanitation lies more in the techniques employed, particularly the way in which pictures depicting typical situations are used. These techniques can and have been adopted for hygiene education activities aimed at students.

Children's Hygiene and Sanitation Training

One interesting example of PHAST approach applications is Children's Hygiene and Sanitation Training (CHAST), which seeks the active involvement of children in efforts to improve their personal hygiene. The CHAST approach is built on the recognition that children are influenced less by abstract concepts than by simple illustrations and examples of the concepts to be taught. The approach uses group work, presentations, songs, and games to develop and share ideas, as well as practical activities, including handwashing, teeth brushing, covering food, and using the toilet. It was designed to be adapted and integrated into the school curriculum.[22]

Like the PHAST approach, the CHAST exercise involves a series of steps:

(i)   An introductory step is to ensure that children become familiar with the facilitators and the methods that they use.

(ii)   During problem identification, cards are used to help children to identify good and bad habits, with particular emphasis on behaviors that can result in the spread of diseases.

(iii) The problem analysis step uses cards to allow children to review the lessons learned during problem identification and then moves on to explanations as to how germs are spread. To make these explanations more accessible to children, posters and guided role plays by some of the children are used to tell "stories" about germ transmission.

(iv)   The practical demonstration step involves practicing of actions to stop the spread of germs. It could include the following exercises: handwashing, toilet use, good practice in water collection and use, tooth brushing, and food handling.

(v)   Results are assessment based on comparison of surveys of behavior carried out before and after the CHAST intervention.

During the problem identification and analysis steps, children are encouraged to work in pairs or in small groups, and then to present their thoughts and findings to larger groups.

---

[22]   Further information can be found in de Vreede (2004), International Reference Centre for Community Water Supply and Sanitation (IRC) and Water Supply and Sanitation Collaborative Council (2004) and CARITAS (2010).

The CHAST activities are meant to be enjoyed, and therefore involve games, exercises, and role plays that prompt children to discuss and genuinely understand the key issues related to personal cleanliness and hygiene. Some CHAST tools include the following:

(i)   **Colored posters.** These are used to introduce characters and situations in stories, show examples of good and bad behavior (to be used in two pile-sorting exercises), and illustrate stories told by children about hygiene-related problems and solutions.

(ii) **Puppets.** These are used in puppet shows to enact WASH-related stories. Practitioners have found puppets particularly useful in engaging young girls and quiet children, who might otherwise be excluded from hygiene promotion activities.

(iii) **Puppet shows.** Puppet shows encourage children to follow and take part in the "antics" of a talking puppet. Puppet shows have proved to be useful in introducing sensitive subjects and allowing discussion of issues that might otherwise be viewed as sacrosanct and not suitable subjects for discussion. Puppet shows should involve humor, which is a good vehicle for encouraging children to discuss sensitive subjects and to make their views known in an open and honest way.

(iv) **Role plays.** Role plays can be used to illustrate situations that are common in everyday life, to raise awareness about common hygiene problems, and to create a positive environment for the discussion of sensitive issues.

(v)   **Card games.** These are designed to reinforce lessons about what constitutes good and bad hygiene behavior. Younger children are encouraged to memorize cards, which is a good way of ensuring that they remember key messages. Older children play a game called "pass the buck," which is used to encourage them to find two cards encouraging good and bad WASH-related practices.

The CHAST exercise is simpler than the PHAST approach and has a strong focus on the hygiene aspects of WASH, while using similar participatory methods. Facilitator training, similar to that required for PHAST, is required for the CHAST facilitators but with a greater emphasis on the learning requirements of children.

A useful manual on the use of the CHAST approach is available from CARITAS (Lowe 2012), which draws on the original experience in Somaliland and subsequent experiences elsewhere in Africa. The manual also provides information on forming and running school health clubs and references to activities and tools not covered by the list provided, including songs and storytelling; specially designed flipcharts that allow teachers to show the front side to groups of children, while consulting the back side of the chart for the script; and suggested discussion questions, games and answers; drawings for coloring (to be used with young children); and pocket charts. Detailed guidelines for facilitators are also included.

The CHAST approach, as developed in sub-Saharan Africa, provides a good starting point for developing improved hygiene education methods for Mongolia, adapted for use in the local context. Its simple methods have been designed specifically for use with children, and with focus on a limited range of key behaviors. The standard CHAST approach uses three characters who appear in stories, role plays, and puppet shows, and have been created to encourage children to discuss hygiene and sanitation-related topics. Careful design is required to ensure that these characters are "recognizable" to children. Thus, it is necessary to create new characters based on Mongolian lifestyles and norms.

Students' Health Clubs

Students can be active agents of hygiene education when they are guided and supported by teachers and school doctors as trained hygiene promotors, who in turn are guided and supported by skilled professionals. Students' hygiene clubs can provide a forum for identifying peer (student) educators and developing their knowledge and skills to carry out hygiene education activities.

This approach was trialed under the UNICEF–AusAID-supported project in Khuvsgul Aimag. School teachers and doctors received training in hygiene promotion and education at the *aimag* center and provided training for students from various clubs organized by child development centers at schools in 12 *soums*. Student hygiene promotors and educators then conducted training for their classmates and lower grade students for about an hour after school. The training covered a range of hygiene topics and was conducted at intervals of no more than 3 months. Doctors, nurses, and midwives from *aimag* healthcare institutions and hospitals provided professional support. The emphasis on regular training rather than one-off training events is important, as regular training leads to a higher likelihood that hygiene promotion lessons are internalized by children, and result in permanent changes in hygiene practice.

APPENDIX 6
Cost Estimation of Improved Water, Sanitation, and Hygiene in Schools

A t the planning stage, it is important to analyze options in terms of their costs and benefits, and determine whether their costs are affordable to schools and local institutions responsible for implementing and managing water, sanitation, and hygiene (WASH) services. Among the questions to be answered are

(i) How much will the proposed WASH facilities cost to build and operate?
(ii) What is the cheapest option for providing the required WASH services?
(iii) Are the costs associated with the preferred option affordable to those who will pay for them?

Appendix 6 provides guidance on assessing and comparing the costs of different options for WASH improvements and determining the affordability of those options. It emphasizes the need to ensure that ongoing operational costs are affordable to the schools. This appendix is intended for professional engineers, architects, and cost estimators who are responsible for planning and designing school WASH, as well as school management, *aimag* and Ulaanbaatar education departments, and the Government of Mongolia, in particular, the Ministry of Education, Culture, Science and Sports (MECSS), and the Ministry of Finance.

Capital and Recurrent Costs

Capital costs are the costs incurred to provide the physical infrastructure required for the delivery of WASH services. Capital costs are made up largely of construction costs but also include

(i) planning and design costs,
(ii) construction supervision costs, and
(iii) equipment and facility replacement costs.

Planning and design costs are incurred prior to construction. For cost estimation purposes, they are usually taken as a proportion, typically 5%–7.5%, of the construction cost, although it may be appropriate to assume a higher proportion for complex and innovative schemes.

The cost of supervision is likely to be in the range of 3%–5% of the construction cost. These figures are approximate and should be refined, with information on planning, design, and construction costs from similar recent projects in Mongolia. Both construction and supervision costs are likely to be higher in remote rural *soums* and *baghs* than in Ulaanbaatar.

Allowance must be made for the fact that facilities and equipment wear out and eventually have to be replaced. Replacement costs are incurred at different time intervals, depending on the item to be replaced. As a general rule, mechanical equipment, including pumps and mechanical treatment units, have to be replaced at more frequent intervals than civil structures such as septic tanks. For cost estimation purposes, the design life of mechanical equipment should typically be taken as 10–15 years, and civil works, 25–30 years.

Recurrent costs include the labor, materials equipment, and power costs incurred in routine operation and maintenance (O&M) of WASH facilities, normally those required within 1 budget year. The cost of hygiene education activities is primarily recurrent.

The construction cost of school WASH facilities, together with planning, design, and construction supervision costs, is normally covered from the central government budget. Recurrent operational costs are normally covered from the *aimag* or Ulaanbaatar education budget, with financial support provided by the central government. This division of responsibilities is important in determining the affordability of different options. In some cases, an option that has the lowest overall cost may prove unsustainable because its annual recurrent costs are higher than the *aimag* education budget. For this reason, the cost analysis of WASH options should include:

(i)   least cost analysis, based on the estimated present values of viable options; and
(ii)   assessment of the recurrent costs of viable options against the WASH O&M budget that is likely to be available.

The objective of the analysis should be to identify the least cost option with recurrent costs that are affordable to schools and *aimag* or Ulaanbaatar education departments that are responsible for the day-to-day management of WASH facilities and services.

Cost Estimation

The starting point for financial appraisal is to make appropriately credible and reliable estimates of capital and recurrent costs. The term "appropriate" is necessary since the required credibility and reliability of estimates are different at various stages in the project cycle, as summarized in Table A6.1.

Table A6.1: Cost Estimation Requirements at Different Stages in the Project Cycle

Stage in Project Cycle	Construction Cost Assessment	Recurrent Cost Assessment
Project identification	"Ballpark" estimate, drawing on information on costs of similar projects	Approximate assessment, with particular focus on power costs and potential costs of equipment replacement
Project concept	Refinement of estimates prepared during project identification, taking into account local factors, particularly cost of delivering water to the school	
Project preparation	Based on quantities of capital goods estimated in the course of preliminary design	Based on preliminary analysis of power, materials, equipment, and labor requirements carried out in the course of preliminary design
Detailed design	Detailed itemized estimates, based on detailed bills of quantities and quoted manufacturer prices	Detailed estimates, based on detailed analysis of power, materials, equipment, and labor requirements

Source: Authors.

Detailed cost estimates should be prepared in local currency units and converted into United States dollars or another appropriate currency as required. For standard items in Mongolia such as pit latrines, cost information is available. [23] It is worthwhile to develop a database of costs for other standard items, including

(i)   borewell installations, including cost of well, pump house, pump, and pipework;
(ii)   water mains of various diameters (cost per 10-meter [m] or 100-m length including fittings);
(iii)   ground level and elevated water storage tanks (with estimated cost per cubic meter [m^3] capacity);
(iv)   standard sanitation block designs, including all internal plumbing;
(v)   septic tanks of various capacities;
(vi)   sewers (cost per 10-m or 100-m length, including manholes and chambers); and
(vii) vehicles (including carts) for use with schemes that involve water transport to schools without a piped water supply.

Where a WASH scheme requires nonstandard facilities, for which there are unlikely to be any readily comparable existing facilities to obtain cost information, a preliminary design exercise is needed at the project concept stage to establish requirements and costs for each of the options identified that are potentially technically feasible. The most credible and reliable approach to estimating construction costs is to multiply estimated quantities by unit rates taken from either a standard schedule of rates (if available); or the rates quoted by contractors for recently awarded contracts. This approach requires that a preliminary design be prepared for each option under consideration with sufficient details so that facility sizes and capacities can be calculated.

One option for estimating costs of frequently required items is to produce a booklet of standard designs, for example, with two pages devoted to each item. Drawings showing details of the item, for instance, a septic tank, should be provided on the left-hand page, while the right-hand page should provide a bill of quantities for the item. The standard designs should take account of the need to protect against freezing, by burying facilities beneath the depth of frost penetration, providing insulation, or some combination of the two.

Recurrent costs include the following:

(i)   energy costs for pumped water supply schemes;
(ii)   energy costs for mechanically aerated wastewater treatment facilities;
(iii)   costs of consumables, including materials such as calcium hypochlorite, which are needed for treatment processes and items such as pump and valve seals that require regular replacement; and items such as paper and markers that are necessary for hygiene education activities; and
(iv)   labor costs, including those incurred to operate and maintain facilities, and implement hygiene education activities.

If there is no existing piped supply or funds are not immediately available to meet the cost of providing a piped supply, water transport in vehicles will be the only short-term option. The operational cost of such water transport should be considered as recurrent costs. As the cost is likely to be high, water transport should be considered only when there is no alternative.

---

[23]   In 2018, the cost of a block of latrines to serve 240 students was typically 90 million togrog (MNT) or about $36,500. Based on the required standard of one toilet for 30–40 users, this price is likely to be for a 6–8 compartment toilet, suggesting that the cost per toilet compartment is in the range of MNT11.25 million to MNT15 million ($4,500–$6,000).

For systems that rely on mechanical equipment to either pump or treat water and/or wastewater, the biggest recurrent costs are often those of power and replacement parts, which may have to be imported. Box A6.1 sets out the procedure for calculating the likely power requirements for pumping.

Box A6.1: Calculation of Energy Cost for Water and Wastewater Pumping

The first step is to calculate the energy requirement for pumps, which depends on the power required and the time during which pumps operate. The basic equation for calculating the power required is:

$$P_h = q \, \rho \, g \, h \, / \, (3.6 \times 106) \tag{6.6.1}$$

where

P_h = hydraulic power (kilowatts [kW])

q = flow capacity (cubic meter per hour [m³/h])

ρ = density of fluid (kilogram per cubic meter [kg/m³]) – 1,000 kg/m³ for water and wastewater

g = acceleration due to gravity (9.81 meter per second squared [m/s²])

h = differential head (meter [m]), including both the static head and head losses due to friction.

To pump 10 m³/h of water against a total head of 15 m:

$$P_h = (10 \text{ m}^3/\text{h}) \, (1000 \text{ kg/m}^3) \, (9.81 \text{ m/s}^2) \, (15 \text{ m}) \, /(3.6 \times 106) = 0.41 \text{ kW} \tag{6.1.2}$$

The amount is the power delivered by the pump. Because of efficiency losses, the power transferred from the motor to the pump (the shaft power) must be higher. To accommodate the need, the hydraulic power must be divided by the efficiency of the pump, which is typically around 0.65 (it is advisable to check the efficiency with the pump manufacturer). Further allowance must be made for the efficiency of the power source for the pump. Electrically operated motors typically run at an efficiency of around 90% when charged at between 50% and 100% of their design load.

Taking account of the pump and motor efficiencies, the power required to pump 15 m³/h against a 10 m head is 0.41/(0.65x0.9) = 0.7 kW.

The total energy requirement for pumping will depend on the period for which the pump is operated. For a 10-hour operating period, the energy requirement will be 10 x 0.7 = 7 kilowatt-hour (kWh).

Where the school has access to a centralized electricity supply system, the total energy cost for the pump can be obtained by multiplying the calculated energy requirement by the appropriate tariff.

For schools without access to a centralized electricity network, pumps are normally powered by either a diesel generator or by diesel-powered pump motors. The operating cost of these systems is normally higher than that of systems that take energy from a centralized electricity system, particularly where fuel has to be transported over a large distance. The fuel required to generate 1 kWh depends on the size of the generator and whether or not it is fully loaded. For the example given above, the fuel requirement is unlikely to be less than 1 liter of diesel per kWh.

Source: Authors.

Comparing Costs

1. Simple Cost Comparison and Its Limitations

The challenge for cost analysis is to compare costs that are incurred at different times. Construction costs are normally incurred at the start of a project while O&M costs continue over the whole life of a facility. Additional costs will be incurred in the future to replace or repair old and faulty facilities and equipment. Different options incur expenditure at different times and with different balances between capital and recurrent expenditures.

Take, for instance, the case of a water supply to serve a school in a rural location in which the options are to supply water by gravity from a surface water source at some distance from the school, or by a pumped supply from local borewells. Let us assume that the capital and operational costs of the two schemes have been calculated, with the results summarized in Table A6.2.

Table A6.2: Summary of Costs for Hypothetical School Water Supply Schemes

	Gravity Scheme	Borewell Scheme
Capital cost	$1 million	$200,000
Annual O&M cost	$20,000	$50,000
New pumps after 15 years		$30,000
Total cost over a 30-year life of scheme	$1.6 million	$2 million

O&M = operation and maintenance.
Source: Authors.

The figures in the bottom row are calculated by adding all the costs incurred over the 30-year life of the schemes. They appear to show that the overall cost of the gravity scheme is cheaper than that of the borewell scheme. However, the figures shown are based on the assumption that a dollar spent today is equivalent to a dollar spent after 10, 20, or 30 years. In fact, a sum of money invested today has a greater value than the same sum invested in the future. The following section introduces the theory that underpins this concept and examines its implications for comparative analysis of the costs of different options over the lifespan of the schemes.

2. Net Present Value

The key concept used in comparison of the costs of different options is that of the time value of money. A sum of money received today can be invested or deposited in a bank, where it will earn interest over the years so that the sum that will be withdrawn after 30 years is higher than the sum invested today. The equation governing this relationship is:

$$P_n = P_0 \times (1 + i)^n \qquad\qquad (1)$$

where

P_0 = initial principal (the amount invested) today

P_n = principle after n years

i = interest rate

The term $(1 + i)^n$ takes account of the fact that when interest is received and added to the principle, the principal increases so that the interest received in the next year will be higher. For an interest rate of 4%, an investment of $100 will increase to about $324 after 30 years.

The equation can be inverted and written in the form:

$$P_0 = P_n/(1+i)^n \qquad (2)$$

where P_0 is the present value of an expenditure P_n incurred after n years. In this equation, the future expenditure P_n has been "discounted" back to the sum P_0 that would have to be invested to realize a sum of P_n after n years. The term $1/(1+i)^n$ is known as the discount factor.

Costs and revenues are normally based on those prevailing at the time that the project is implemented. It is then necessary to make a distinction between the nominal rate, which is based on the interest rate charged by commercial banks, and the real rate, which takes into account inflation. The real rate is given by the equation:

$$r = (i-h)/(1+i) \qquad (3)$$

where r = real interest rate

i = bank interest rate

h = inflation rate

Thus, for example, if the interest rate is 10% and the inflation rate is 5%, the real discount rate is given by (0.1-0.05)/(1+0.05), which equals 0.05/1.05 or 0.0476 (4.76%). It can be seen that this figure approximates to the interest rate minus the inflation rate.

When assessed in terms of current prices, recurrent operational costs tend to remain constant from year to year. The present value of a series of constant costs running from year 1 to year n is obtained by multiplying the annual cost by the expression $[(1+i)^n -1]/[i(1+i)^n]$ to give the following equation:

$$P_{rs} = RC \times [(1+i)^n -1]/[i(1+i)^n] \qquad (4)$$

where P_{rs} is the present value of an annual recurrent cost RC incurred from year 1 to year n inclusive. The term $[(1+i)^n -1]/[i(1+i)^n]$ is referred to as the uniform series present worth (USPW) factor. Discount factors and USPW factors can be calculated from first principles but are also available in standard tables.

Because USPW factors are obtained by adding discounted costs, it is possible to subtract years when there is no expenditure. For instance, if O&M costs start in year 2 and continue to year 20, the present value of these costs is obtained by multiplying the annual cost by the USPW factor for 20 years less the USPW factor for 1 year, i.e.,

$$\text{Present value} = [USPW(20)-USPW(1)] \times \text{Annual cost} \qquad (5)$$

Discounting always reduces the contribution of future expenditure, especially with higher discount rates. For this reason, a high discount rate favors options with relatively low capital and high recurrent costs, particularly when costs are discounted over a long period, say 30 years. Conversely, a low discount rate favors options with relatively high capital and low recurrent costs. The discount rate used for analysis is thus important.

Standard financial appraisal practice uses the prevailing private discount rate, which reflects the opportunity cost of capital and may exceed 6%. When evaluating policies and decisions relating to public goods such as education, it will be more appropriate to use the social discount rate. Social discount rate is briefly explained in Box A6.2. In the absence of relevant local information, a social discount rate in the range 1. 5%–3.5% can be assumed.

Box A6.2: Brief Introduction to the Concept of the Social Discount Rate

The social discount rate can be defined as the social rate of time preference, in essence, representing the rate at which the society would trade a unit of benefit between the present and the future. The social rate of time preference has two components:

(i)   the rate at which individuals discount future consumption over present consumption, on the assumption that no change in per capita consumption is expected; and

(ii) an additional element, to allow for growth in per capita consumption over time, which implies that future consumption will be higher relative than present consumption and thus have lower marginal utility.

Taking these two components together, the social discount rate can be represented by the equation:

$$r = \rho + \mu^* g \qquad\qquad (6.2.1)$$

where   r = social preference rate (the social discount rate)

ρ = first component

μ = elasticity of marginal utility of consumption (μ) with respect to utility

g = annual growth in per capita consumption

Assuming annual growth in per real capita consumption, g, of 2%, an assumed rate of time preference (ρ) equal to 1.5%, and an elasticity of marginal utility of consumption μ equal to 1, the current recommended social discount rate is

$$r = 0.015 + 1.0^* 0.02 = 3.5\% \qquad\qquad (6.6.2)$$

Source: WERF, Water Research Foundation, GWRC, and GHD Consulting Inc.
http://simple.werf.org/simple/media/documents/BCT/stepFourLinks/01A.html.

Table A6.3 sets out the discounted costs for the school water supply options presented in Table A6.2. The capital costs are incurred immediately and thus do not require discounting. The O&M cost for the whole 30 years has been discounted using the USPW factor, and the pump replacement cost after 15 years has been discounted, which have then been included in the net present value. The replacement cost after 30 years has not been included in the calculation as the assumption is that both schemes need to be replaced after 30 years.

Table A6.3: Discounted Costs for Gravity and Borewell Schemes
($)

	Gravity Scheme		Borewell Scheme	
Discount rate (%)	2	5	2	5
Capital cost	1	1	0.2	0.2
Discounted O&M cost	0.448	0.307	1.120	0.769
Pump replacement cost	0.000	0.000	0.017	0.007
Net present value	1.448	1.307	1.336	0.976

O&M = operation and maintenance.
Source: Authors.

Based on Tables A6.2 and A6.3, the net present values of the two schemes are in Table A6.4.

Table A6.4: Net Present Value for Gravity and Borewell Schemes
($)

	Gravity Scheme	Borewell Scheme
Without discounting	1.60	2.00
Applying 2% discount rate	1.45	1.34
Applying 5% discount rate	1.31	0.98

Source: Authors.

These tables show that applying a discount rate reduces the net present value of the borewell scheme, relative to that of the gravity scheme, so that at both 2% and 5% discount rates, the borewell scheme appears to be the better option. This analysis illustrates the point that higher discount rates tend to favor schemes with low capital expenditure and high recurrent expenditure. However, as already noted, an option is only viable if its recurrent costs can be met from the budget of the organization that is responsible for ongoing management, operation, and maintenance of facilities and services. This suggests that while the borewell scheme has the lower net present value, the gravity scheme is still the best option if the school does not have access to the funds required to meet its higher O&M costs.

Glossary

aimag	an administrative unit equivalent to a province in Mongolia
baffled reactor	anaerobic wastewater treatment technology consisting of initial settler compartment followed by series of upward-flow compartments
bagh	subunit of a *soum*
black water	wastewater from flushed toilets, containing feces, urine and flush water
borehole or borewell	well constructed by boring a vertical hole, which is normally subsequently fitted with a casing
ceramic filter	a filtering device that uses a porous ceramic material as the filtering agent
coagulation	grouping together of small particles in a solution into larger particles
constructed wetland	simple aerobic treatment technology that aims to replicate the processes that occur in natural wetlands under controlled conditions
container-based sanitation	sanitation system in which small containers located below toilets are designed to be frequently removed, transported to, and emptied at a treatment facility
drainfield	system of perforated pipes surrounded by gravel, provided to allow effluent from a septic tank or other form of treatment to percolate into the ground
dry direct-drop toilet	toilet in which excreta are deposited directly through a hole into a pit or vault
dry sanitation	sanitation system that does not require water for flushing
effluent	wastewater-treated effluent is the wastewater discharged after treatment in a wastewater treatment plant
Escherichia coli (E. coli)	thermotolerant coliform bacteria, an indicator of human fecal contamination
excreta	human wastes, including urine and feces
fecal-oral disease	disease transmitted from feces of infected person to mouth of another person
fecal sludge	waste from toilets and other sanitation facilities that accumulates in septic tanks, pits and vaults
filtration	process of separating suspended solid matter from a liquid, by causing the liquid to pass through the pores of a membrane
ger	a traditional tent where seminomadic herders live

gravity supply	supply of water from a higher to a lower point so as to eliminate the need for pumping
gray water	alternative name for sullage; wastewater from kitchens and bathrooms, excluding excreta
groundwater	water present beneath the earth's surface in soil pore spaces and in fractures of rock formations
hygiene	practices designed to keep a person and his or her surroundings free of pathogens in order to prevent the spread of diseases
infiltration gallery	structure consisting of a network of perforated pipes surrounded by gravel, located in the bed of a water body (lake, stream or river) and used to collect sub-surface water located below the water body
intake	structure designed to allow water to be transferred from a surface water source to the water delivery system
lagoon	shallow, engineered pond used for storage and treatment of either wastewater or fecal sludge; when wastewater is treated, lagoon is another name for waste stabilization pond
leakage	water loss from leaks in the distribution system and house connections
looped water supply system	water distribution system in which water points connect to mains that interconnect to form loops rather than dead-end branches
moving bed biological reactor	form of aerobic sewage treatment that combines aeration with attached growth microbiological processes on plastic media
nanofiltration	filtration in which a substance is put through a filter with holes that are measured in nanometers
package plant	a pre-manufactured treatment facility, used to treat wastewater from a small community or individual property
peak factor	ratio of peak flow to average flow; peak day or peak hour flow in the case of water supply
permafrost	subsoil that is permanently frozen
photovoltaic solar panels	panels made up of solar cells that convert energy from the sun into electricity
pit latrine	direct-drop sanitation facility, in which excreta are deposited into a pit
pitless adapter	device that allows passage of well piping out through the side of a well casing at a point underground; pitless adapters eliminate the need for a well pit to protect pipes from freezing
polyethylene	polymer of ethylene, widely used as a pipe material
pour-flush toilet	toilet from which excreta are flushed using water poured by hand from a suitable container
rotating biological contactor	aerobic treatment technology that relies on microbiological processes on attached growth on slowly circulating discs, half submerged in wastewater

"R" value	a measure of the ability of a material to resist the flow of heat from its hot side to its cold side; the higher the "R" value, the higher the insulation value of the material
sanitation	system for disposal of wastes, including excreta and liquid wastes; often used to include solid waste management
SATO pan	toilet pan with a hinged flap that opens under the weight of excreta and closes automatically once the excreta have fallen into the pit, vault or container below
septic tank	tank, normally with two compartments, which treats wastewater by sedimentation and anaerobic digestion
sewage	wastes from houses, factories, businesses, and institutions, consisting of a mixture of excreta, wash water, and wastewater from kitchens and bathrooms
sewerage	system of pipes to convey sewage
soakaway	pit, either filled with rocks or with a permeable lining, from which wastewater percolates into the ground
sorptive filtration	filtration preceded by use of suspended particulate matter to absorb impurities
soum	subunit of an aimag
sullage	wastewater from kitchens and bathrooms, not including excreta
surface water	water found on the surface of the earth in streams, rivers, and lakes
upward-flow anaerobic filter	wastewater treatment technology in which anaerobic wastewater flows upwards through a sand or gravel filter
uPVC	unplasticized polyvinyl chloride, a rigid, chemically resistant form of PVC (polyvinyl chloride)
user interface	means by which users access water and dispose of waste products
vacuum tanker	truck mounted with cylindrical holding tank, into which liquid wastes from septic tanks and wet pits can be drawn using a vacuum pump
vault toilet	dry direct-drop toilet in which excreta are held in an above-ground compartment
ventilated improved pit latrine	improved form of pit latrine, with dark interior and vertical pipe, topped with a screen, designed to vent the pit
wastage	water losses through leaking appliances, taps left open and overflowing water tanks
waste stabilization pond	sewage treatment technology that relies on natural processes occurring in ponds of various depths; ponds may be designed to be anaerobic, facultative (part aerobic and part anaerobic), or wholly aerobic
water point	means by which consumer accesses water, normally a tap
water storage	storage provided within the water supply system

References

Adams J., J. Bartram, Y. Chartier, and J. Sims (eds.). 2009. *Water, Sanitation and Hygiene Standards for Schools in Low-cost Settings*. Geneva: World Health Organization. http://www.who.int/water_sanitation_health/publications/wash_standards_school.pdf.

ADB. 2014. *Demand in the Desert: Mongolia's Water–Energy–Mining Nexus*. Manila.

Alberta Agriculture and Forestry. 2012. Agri-facts, Pitless Adaptors, Agdex716(C29).

Alcanz Consulting Group. 2015. *Final Evaluation Report: WASH in Schools and Kindergartens Project*. New York: UNICEF. https://www.unicef.org/evaldatabase/files/Mongolia_FINAL_Evaluation_Report_WASH_in_Schools_and_Kindergartens.pdf.

Altanzagas, B. 2006. Report of the Project Healthy Springs in Mongolia. Ulaanbaatar: Government of Mongolia and WHO/AGFUND.

Alter, A. J. 1969. Water Supply in Cold Regions. *Cold Regions Science and Engineering Monograph*. III–C5a. New Hampshire: US Army Cold Regions Research and Engineering Laboratory.

Associação Brasileira de Normas Técnicas. 1982. Construcao e instalacao de fossas septicas e disposicao dos efluentes finais. Rio de Janeiro (NBR 7229).

AusAID. 2011. *Mongolia: WASH in Schools and Kindergartens: Project Design Document*. http://washinschoolsmapping.com/wengine/wp-content/uploads/2015/10/mongolia-wash-design-document.pdf.

Bathold, K, Z. Tuul, and B. Oyun. 2004. Access to Water and Sanitation Services in Mongolia. Ulaanbaatar: Government of Mongolia, UNDP, UNICEF, WHO.

Biran, A, A. Tabyshalieva, and Z. Salmorbekova. 2005. Formative Research for Hygiene Promotion in Kyrgyzstan. *Health Policy and Planning*. 20. pp. 213–221.

Biran, A., W. P. Schmidt, L. Zeleke, H. Emukule, H. Khay, J. Parker, and D. Peprah. Hygiene and Sanitation Practices among Residents of Three Long-Term Refugee Camps in Thailand, Ethiopia and Kenya. *Tropical Medicine and International Health*. 17 (9). pp. 1133–1141. http://onlinelibrary.wiley.com/doi/10.1111/j.1365-3156.2012.03045.x/pdf.

Bobrick. 2017. *Planning Guide for Accessible Restrooms*. http://www.bobrick.com/documents/planningguide.pdf.

Bohnert, K., A. Chard, A. Mwaki, A. Kirby, R. Muga, C. Nagel, E. Thomas, and M. Freeman. 2016. Comparing Sanitation Delivery Modalities in Urban Informal Settlement Schools: A Randomized Trial in Nairobi, Kenya. *International Journal of Environmental Research and Public Health* 13, 1189. http://www.communityledtotalsanitation.org/sites/communityledtotalsanitation.org/files/Sanitation_Delivery_Modalities_Urban_Schools.pdf.

Bräuner, E. V., R. B. Nordsborg, Z. J. Andersen, A. Tjønneland, S. Loft, and O. Raaschou-Nielsen. 2014. Long-Term Exposure to Low-Level Arsenic in Drinking Water and Diabetes Incidence: A Prospective Study of the Diet, Cancer and Health Cohort. *Environment Health Perspectives.* 122 (10). http://ehp.niehs.nih.gov/1408198/.

Buttle, M. and M. Smith. 2004. *Out in the Cold: Emergency Water Supply and Sanitation for Cold Regions.* 3rd ed. Leicestershire, UK: Water, Engineering and Development Centre, Loughborough University. http://wedc.lboro.ac.uk/resources/books/Out_in_the_Cold_-_Complete.pdf.

CARITAS. 2010. AGUASAN Meeting n°106. http://www.sswm.info/sites/default/files/reference_attachments/CARITAS%202010%20CHAST%20Presentation.pdf.

Centers for Disease Control and Prevention. Ceramic Filtration. http://www.cdc.gov/safewater/ceramic-filtration.html.

Cold Climate Research Centre. n.d. *The Alaska Consumer Guide to Home Heating.* Anchorage: Alaska Housing Finance Corporation.

Cunningham, W. L. and C. W. Schalk, comps. 2011. Groundwater Technical Procedures of the US Geological Survey. *US Geological Survey Techniques and Methods.* No. 1–A1. https://pubs.usgs.gov/tm/1a1/.

Curtis, V., L. Danquah, and R. Aunger. 2009. Planned, Motivated and Habitual Hygiene Behaviour: An Eleven Country Review. *Journal of Health Education Research.* 24 (4). pp. 655–673. https://www.ncbi.nlm.nih.gov/pmc/articles/PMC2706491/.

de Vreede, E. 2004. CHAST Children Hygiene and Sanitation Training in Somalia. In M. Snel, K. Shordt, and A. Mooijman, *Proceedings of School Sanitation and Hygiene Education Symposium.* Delft, Netherlands. http://www.sswm.info/content/child-hygiene-and-sanitation-training-chast.

Department of Education and Skills. 2014. Guidelines and Standards for Sanitary Facilities in Primary Schools. *Technical Guidance Document.* No. TGD–021-2. http://www.education.ie/en/School-Design/Technical-Guidance-Documents/Current-Technical-Guidance/TGD%E2%80%93021-2-Guidelines-and-Standards-for-Sanitary-Facilities-in-Primary-Schools-1st-Edition-April-2014-.pdf.

DODEA. *Design Guidelines—Mounting Heights.* http://www.dodea.edu/edSpecs/upload/5_Design_Building_Mounting.pdf.

Dudley, E. 1993. *The Critical Villager: Beyond Community Participation.* London and New York: Routledge.

E3Design Ltd. Ghanasan Project. http://e3design.co.uk/ghanasan--clean-team-toilet.html.

Emory University and UNICEF. n.d. Equity of Access to WASH in Schools: A Comparative Study of Policy and Service Delivery in Kyrgyzstan, Malawi, the Philippines, Timor-Leste, Uganda and Uzbekistan. New York: UNICEF. https://www.unicef.org/wash/schools/files/Equity_of_Access_to_WASH_in_Schools.pdf.

EOSS and Water, Engineering and Development Centre. 2014. *A Collection of Contemporary Toilet Designs.* Leicestershire, UK: Water, Engineering and Development Centre, Loughborough University. http://wedc. lboro.ac.uk/resources/books/Contemporary_Toilet_Designs.pdf.

Erhard, L., J. Degabriele, D. Naughton, and M. C. Freeman. 2013. Policy and Provision of WASH in Schools for Children with Disabilities: A Case Study in Malawi and Uganda. *Global Public Health.* 8 (9). 1000–1013.

Food and Agriculture Organization of the United Nations (FAO). 1996. Simple Methods for Aquaculture: Management for Freshwater Fish Culture: Ponds and Water Practices. *Training Series.* No. 21/1. Rome.

Greywater Action. Greywater Systems in Freezing Climates. http://greywateraction.org/systems-for-cold-climates-including-wetlands/.

Hanna, T. M., E. Sundberg, and G. Hayden. 2016. Installation of Infiltration Gallery at Greens Creek Mine – Juneau, Alaska. Universal Journal of Geoscience 4(6): 122-127. DOI: 10.13189/ujg.2016.040602.

Hydromatch. *Flow Estimation for Streams and Small Rivers.* http://www.hydromatch.com/sites/default/files/downloads/DIY-flow-measurement-guide.pdf.

Ilias J., and B. Scharaw n.d. MOMO Fact Sheet: Decentralized Waste Water Management: Experiences from Pilot Operation in Orkhon Sum, MOMO Integrated Water Resources Management. http://iwrm-momo.de/download/MoMo%20Fact%20Sheet%20WWTP%20Orkhon%20Sum.pdf.

International Reference Centre for Community Water Supply and Sanitation (IRC). 2006. *Children's Health Clubs in Schools—Opportunities and Risks.* http://www.ehproject.org/PDF/ehkm/irc-health_clubs.pdf.

IRC and Water Supply and Sanitation Collaborative Council. 2004. *School Sanitation & Hygiene Education Symposium. The Way Forward: Construction Is Not Enough.* Symposium Proceedings & Framework for Action. https://www.ircwash.org/sites/default/files/Snel-2004-School.pdf

Lowe, E. 2012. *CHAST Children's Hygiene and Sanitation Training: A Practical Facilitation Handbook.* Switzerland, Luxembourg, Nairobi, Kenya: CARITAS. http://www.pseau.org/outils/ouvrages/caritas_chast_children_s_hygiene_and_sanitation_training_2012.pdf.

Lstiburek, J. 2010. BSI-031: Building in Extreme Cold. https://buildingscience.com/documents/insights/bsi-031-building-in-extreme-cold.

Masonry Heater Association of North America. What is a Masonry Heater? http://www.mha-net.org/what-is-a-masonry-heater/.

Ministry of Education, Culture, Science and Sports; Ministry of Health; and Ministry of Finance. 2015. Joint Ministerial Order on Norms and Requirements of Water, Sanitation, and Hygiene for Kindergartens, Secondary Schools, and Dormitories. No: A/253, 251 and 173. Ulaanbaatar.

Monteiro, M. S. and J. R. Gomes. 1998. Reestruturação Produtiva e Saúde do Trabalhador: Um Estudo de Caso. *Cadernos de Saúde Pública.* 14 (2). pp. 345–353. http://www.scielo.br/scielo.php?script=sci_arttext&pid=S0102-311X1998000200011.

National Renewable Energy Laboratory. 2001. *Wind Energy Resource Atlas of Mongolia.* Washington, DC. https://www.nrel.gov/docs/fy01osti/28972.pdf.

National Snow & Ice Data Center. https://nsidc.org/sites/nsidc.org/files/images/data/ggd/geocryo_regions.png.

Patinet, J. and A. Delmaire. 2015. *External Evaluation of ACF's Project Improve Access to Water, Sanitation and Hygiene in the Ger Areas of Ulaanbaatar by Providing Proven and Innovative Solutions.* Plaisians, France.

Peal, A., B. Evans, and C. van der Voorden. 2010. *Hygiene and Sanitation Software: An Overview of Approaches.* Geneva: Water Supply & Sanitation Collaborative Council. http://www.eawag.ch/fileadmin/Domain1/ Abteilungen/sandec/schwerpunkte/sesp/CLUES/Toolbox/t1/D1_2_Peal_2010.pdf.

Pinfold, J. V. 1990. Fecal Contamination of Water and Finger Tip-Rinses as a Method for Evaluating the Effect of Low-Cost Water-Supply and Sanitation Activities on Feco-Oral Disease Transmission. A Case-Study in Rural North-East Thailand. *Epidemiology and Infection.* No. 105. pp. 377–389. https://www.ncbi.nlm.nih.gov/ pmc/articles/PMC2271894/pdf/epidinfect00023-0156.pdf75.

Punsalmaa, B., B. Nyamsuren, and B. Buyndalai. 2004. Trends in River and Lake Ice in Mongolia. *AIACC Working Paper.* No. 4. http://www.start.org/Projects/AIACC_Project/working_papers/Working%20Papers/AIACC_ WP_No004.pdf.

Restroom Association (Singapore). 2018. *A Guide to Better Public Toilet Design and Maintenance.* 4th ed. http:// www.toilet.org.sg/articles/GuideBetterPublicToilet.pdf.

Sanergy. Our Model. http://www.sanergy.com/our-model/

Seifert, R. D. 2004. *Suggestions for Installing Domestic Water Storage Tanks.* Alaska: University of Alaska Fairbanks Cooperative Extension Service. http://www.builditsolar.com/Projects/Water/HCM-04950.pdf.

Soil. https://www.oursoil.org/.

SOLARGIS. Download Solar Resource Maps and GIS Data for 180+ Countries. https://solargis.com/maps-and-gis-data/download/mongolia/.

Structure Now Social Solutions. http://www.structurenow.com/ablution-toilets.php.

Sustainable Sanitation and Water Management Toolbox. https://sswm.info/sswm-solutions-bop-markets/ affordable-wash-services-and-products/affordable-water-supply/arsenic-removal-technologies.

UNICEF and GIZ. 2013. Field Guide: The Three Star Approach for WASH in Schools. New York: United Nations Children's Fund. https://www.unicef.org/wash/schools/files/UNICEF_Field_Guide-3_Star-Guide.pdf.

UNICEF and IRC. 1998. A Manual on School Sanitation and Hygiene, Water, Environment and Sanitation. *Technical Guidelines Series.* No. 5. New York: United Nations Children's Fund. https://www.unicef.org/wash/ files/Sch_e.pdf.

UNICEF and REACH. 2018. *Jordan. WASH in Schools Infrastructure Assessment and KAP Survey. Za'atari and Azraq.* November 2018. https://reliefweb.int/sites/reliefweb.int/files/resources/reach_jor_wins_report_unicef.pdf.

UNICEF. 2013. Improving Child Nutrition: The Achievable Imperative for Global Progress. New York: United Nations Children's Fund. http://data.unicef.org/wp-content/uploads/2015/12/NutritionReport_April2013_ Final_29.pdf.

UNICEF. 2014. *WASH in Schools Distance-Learning Course Learnings from the Field 2014.* New York: United Nations Children's Fund. http://www.unicef.org/wash/schools/files/Learnings_From_the_Field_2014.pdf.

UNICEF Mongolia. 2012. *WASH in Schools and Kindergartens Project. Progress Report (PBA:SC/2012/0100).* Report for the Department of Foreign Affairs and Trade—*AusAID.* Ulaanbaatar.

United States Environmental Protection Agency (US EPA). 1994. *Drinking Water Treatment for Small Communities: A Focus on EPA's Research.* Washington, DC: Office of Research and Development, Environmental Protection Agency. http://infohouse.p2ric.org/ref/02/01998.pdf.

University of Alaska Fairbanks Cooperative Extension Service. n.d. *Passive Solar Heating: An Energy Factsheet.*

Wagner, E. G. and J. N. Lanoix. 1958. Excreta Disposal for Rural Areas and Small Communities. *World Health Organization Monograph Series* No. 39. Geneva: World Health Organization. https://apps.who.int/iris/handle/10665/41697.

Washington State Department of Health. 2017. *Preventive Maintenance Program—Guide for Small Public Water Systems Using Groundwater.* http://www.doh.wa.gov/portals/1/documents/pubs/331-351.pdf.

Water Systems Council. 2013. Pitless Adapters, Pitless Units and Well Caps. Washington, DC: Water Systems Council National Programs Office.

WERF, Water Research Foundation, GWRC, and GHD Consulting Inc. http://simple.werf.org/simple/media/documents/BCT/stepFourLinks/01A.html.

Women in Europe for a Common Future. 2006. *Ecological Sanitation and Hygienic Considerations for Women Fact Sheet.*

World Bank. 2019. Global Solar Atlas 2.0, Solar Resource Data: SOLARGIS.

World Health Organization (WHO). 1997. The Physical School Environment An Essential Component of a Health-Promoting School. *Information Series on School Health.* Document 2. Geneva. http://www.who.int/school_youth_health/media/en/physical_sch_environment_v2.pdf.

WHO. 1998. PHAST Step-by-Step Guide: A Participatory Approach for the Control of Diarrhoeal Diseases. Geneva.

WHO. *Flow Measurement and Control. Fact Sheet 2.9.* http://www.who.int/water_sanitation_health/hygiene/emergencies/fs2_9.pdf.

World Vision. Improved Sanitation—No More Latrine! https://www.wvi.org/mongolia/video/improved-sanitation-no-more-latrine.

WTE Ltd. Tank Legislation and Regulations. https://www.wte-ltd.co.uk/wastewater_legislation.html.